Lecture Notes in Mathematics

W0112269

Editors:
J.-M. Morel, Cachan
B. Teissier, Paris

'

For further volumes:
http://www.springer.com/series/304

FONDAZIONE CIME
ROBERTO CONTI
CENTRO INTERNAZIONALE MATEMATICO ESTIVO
INTERNATIONAL MATHEMATICAL SUMMER CENTER

Fondazione C.I.M.E., Firenze

C.I.M.E. stands for *Centro Internazionale Matematico Estivo*, that is, International Mathematical Summer Centre. Conceived in the early fifties, it was born in 1954 in Florence, Italy, and welcomed by the world mathematical community: it continues successfully, year for year, to this day.

Many mathematicians from all over the world have been involved in a way or another in C.I.M.E.'s activities over the years. The main purpose and mode of functioning of the Centre may be summarised as follows: every year, during the summer, sessions on different themes from pure and applied mathematics are offered by application to mathematicians from all countries. A Session is generally based on three or four main courses given by specialists of international renown, plus a certain number of seminars, and is held in an attractive rural location in Italy.

The aim of a C.I.M.E. session is to bring to the attention of younger researchers the origins, development, and perspectives of some very active branch of mathematical research. The topics of the courses are generally of international resonance. The full immersion atmosphere of the courses and the daily exchange among participants are thus an initiation to international collaboration in mathematical research.

C.I.M.E. Director
Pietro ZECCA
Dipartimento di Energetica "S. Stecco"
Università di Firenze
Via S. Marta, 3
50139 Florence
Italy
e-mail: zecca@unifi.it

C.I.M.E. Secretary
Elvira MASCOLO
Dipartimento di Matematica "U. Dini"
Università di Firenze
viale G.B. Morgagni 67/A
50134 Florence
Italy
e-mail: mascolo@math.unifi.it

For more information see CIME's homepage: http://www.cime.unifi.it

CIME activity is carried out with the collaboration and financial support of:

- INdAM (Istituto Nazionale di Alta Matematica)

- MIUR (Ministero dell'Istruzione, dell'Università e della Ricerca)

- Ente Cassa di Risparmio di Firenze

Luca Capogna • Pengfei Guan
Cristian E. Gutiérrez • Annamaria Montanari

Fully Nonlinear PDEs in Real and Complex Geometry and Optics

Cetraro, Italy 2012

Editors:
Cristian E. Gutiérrez
Ermanno Lanconelli

 Springer

FONDAZIONE
CIME
ROBERTO CONTI

Luca Capogna
Department of Mathematical Sciences
Worcester Polytechnic Institute
Worcester, MA, USA

Pengfei Guan
Mathematics and Statistics
McGill University
Montreal, QC, Canada

Cristian E. Gutiérrez
Department of Mathematics
Temple University
Philadelphia, PA, USA

Annamaria Montanari
Dipto. Matematica
Università di Bologna
Bologna, Italy

ISBN 978-3-319-00941-4   ISBN 978-3-319-00942-1 (eBook)
DOI 10.1007/978-3-319-00942-1
Springer Cham Heidelberg New York Dordrecht London

Lecture Notes in Mathematics ISSN print edition: 0075-8434
                             ISSN electronic edition: 1617-9692

Library of Congress Control Number: 2013945179

Mathematics Subject Classification (2010): 31-02, 32-T-15, 32-02, 35-02, 35-J-70, 35-J-96,
                             53-A-10, 53-02, 78-02, 78-A-05, 78-A-25

Printed on acid-free paper

Springer is part of Springer Science+Business Media (www.springer.com)

Preface

This volume contains the notes from the CIME school "Fully Nonlinear PDEs in Real and Complex Geometry and Optics" held at Cetraro (Cosenza, Italy) during the week of July 9–13, 2012. The school consisted of four courses: *Extremal problems for quasiconformal mappings in space* by Luca Capogna, *Fully nonlinear equations in geometry* by Pengfei Guan, *Monge-Ampère type equations and geometric optics* by Cristian E. Gutiérrez, and *On the Levi Monge–Ampère equation* by Annamaria Montanari.

The purpose of the school was to present current areas of research, arising both in the theoretical and in the applied settings, that involve fully nonlinear partial differential equations. The solution to these problems requires the development of ad hoc techniques arising in the relevant geometrical framework, and in case of the applications, determined by the physical laws governing the studied phenomena. Precisely, the equations presented in the school come from conformal mapping theory, differential geometry, optics, and geometric theory of several complex variables.

The following is a quick overview of the contents of each course.

Luca Capogna's lectures provided an introduction to two vector-valued extremal L^∞-variational problems involving mappings. The first one concerns a classical problem in geometric function theory that first arose in a work of Grotzsch from 1928: among all orientation preserving quasiconformal homeomorphisms $w : \Omega \to \Omega'$ whose traces agree with a given mapping $u_0 : \partial\Omega \to \partial\Omega'$, find one which minimizes the functional $u \to \left\| \dfrac{|du|}{(\det du)^{\frac{1}{n}}} \right\|_\infty$. Variants of this problem occur when, instead of using boundary data, the class of competitors is defined in terms of a fixed homotopy class or by requesting that the traces map quasi-symmetrically $\partial\Omega$ into $\partial\Omega'$. The second problem goes back to two papers from 1934, one by Whitney and the other by MacShane, and leading to the recent theory of *absolutely minimal Lipschitz extensions*. That is, let $\Omega \subset \mathbb{R}^n$ be open, $F \subset \overline{\Omega}$ be a compact set, and let $g \in Lip(F, \mathbb{R}^m)$. Among all Lipschitz extensions of g from F to Ω, is there a canonical unique extension that in some sense has the smallest possible Lipschitz norm? The natural questions arising in connection with these problems are

existence, uniqueness, and structure of the minimizers. The last few decades have
seen intense activity from different communities of mathematicians in the study
of both problems. However, at this time there does not seem to be much synergy
and communication between these communities, both in terms of shared techniques
used in the study of these problems and in terms of common point of views. One
of the goals of Capogna's notes is to foster such synergies by outlining some of the
common features in these problems.

Pengfei Guan's lectures considered nonlinear elliptic and parabolic partial differential equations arising from geometric problems for hypersurfaces in $\mathbb{R}^{n+1}$. The
notes are an introduction to geometric analysis. A curvature-type elliptic equation is
used to solve the problem of prescribing curvature measures, which is a Minkowski-
type problem. Curvature measures are defined using the Steiner formula. An inverse
mean curvature-type parabolic equation is employed for the proof of isoperimetric-
type inequalities for quermassintegrals of k-convex star-shaped domains. Both types
of equations are fully nonlinear geometric PDEs. The emphasis of Guan's notes
is on a priori estimates, a key step in the theory of fully nonlinear PDEs. The
presentation is self-contained and requires basic knowledge of PDEs and geometry,
namely the standard maximum principles for linear elliptic and parabolic equations,
the elementary formulas of Gauss, Codazzi, and Weingarten for hypersurfaces in
$\mathbb{R}^{n+1}$, and the curvature commutator identities. Two basic and deep results are used
without proofs: the Evans–Krylov theorem for uniformly fully nonlinear elliptic
equations and Krylov's theorem for uniformly parabolic fully nonlinear PDEs.

Gutiérrez's course presented an introduction to Monge–Ampère-type equations
and its applications to geometric optics. In general, these equations involve the
Jacobian determinant of a map and arise in the mathematical description of
numerous optical, acoustic, and electromagnetic applications, in particular, in lens
and reflector antenna design. The geometric optics problems considered concern
refraction, reflection, or both. A typical refraction problem that was considered in
the lectures is the following: suppose we have two homogenous media I and II with
different refractive indices, a light beam emanates from a point O, surrounded by
medium I, and we seek an interface surface separating media I and II described by
$\{\rho(x)x : x \in \Omega\}, \Omega \subset \mathbb{S}^{n+1}$, and such that all rays are refracted into either a set of
prescribed directions or illuminate a given object lying on a surface, say on a plane
in medium II. The first type of problem is called far field and the second near field.
These two problems are of different mathematical nature: the first one is variational
and the second is not. The input and output intensities of radiation are prescribed
and so the problem of finding the interface surface is a typical inverse problem.
Gutiérrez's notes explain how to solve these problems: in the far field case using
mass transport, and in the near field case using the Minkowski method. The physical
background underlying these problems is explained using Maxwell equations, and,
as a consequence, a deduction of Fresnel formulas is presented. These formulas
are finally applied to model the case when there is loss of energy due to internal
reflection.

Annamaria Montanari's course focused on the Levi Monge–Ampère equation.
The equation is related to notions of curvatures associated with pseudoconvexity

and the Levi form, in a way similar to how the classical Gauss and mean curvatures are related to the convexity and the Hessian matrix. Given a nonnegative function k, the Levi Monge–Ampère equation for the graph of a function $u : \mathbb{R}^{2n+1} \to \mathbb{R}$ is

$$\det \Lambda = k(x; u)(1 + |Du|^2)^{\frac{n+1}{2}},$$

where Λ is the Levi form of the graph of u and Du is the Euclidean gradient of u. More generally, in her notes, Montanari considers elementary symmetric functions of the eigenvalues of the Levi form Λ and shows that these curvature equations contain geometric information about hypersurface considered. Next, the notes show that the curvature operator leads to a new class of second order fully nonlinear equations whose characteristic form, when computed on generalized pseudoconvex functions, is nonnegative definite with a kernel of dimension one. Thus, the equations are not elliptic at any point. However, they have the following redeeming feature: the missing ellipticity direction can be recovered by suitable commutation relations. Using this property, existence, uniqueness, and regularity of viscosity solutions of the Dirichlet problem for graphs with prescribed Levi curvature are proved. Basic notions and results from the theory of functions of several complex variables, and from the theory of viscosity solutions for fully nonlinear degenerate elliptic equations, are also described in the notes.

It is a great pleasure to thank the speakers for their very interesting lectures and for the useful lectures notes they have prepared for this volume. We also want to thank all the participants of the school for their interest in these subjects.

Finally, we would like to warmly thank the CIME foundation for giving us the opportunity to organize and finance this school. We give our special gratitude to Pietro Zecca, Elvira Mascolo, and all the CIME staff for their invaluable help and support. Also, a special thanks to Mrs. Ute McCrory at Springer for her assistance in preparing this volume.

Worcester, MA   Luca Capogna
QC, Canada   Pengfei Guan
Philadelphia, PA   Cristian E. Gutiérrez
Bologna, Italy   Annamaria Montanari
Bologna, Italy   Ermanno Lanconelli
April 5, 2013

Contents

Contributors

Luca Capogna Department of Mathematical Sciences, Worcester Polytechnic Institute, Worcester, MA, USA

Institute for Mathematics and Its Applications, University of Minnesota, Minneapolis, MN, USA

Pengfei Guan Department of Mathematics, McGill University, Montreal, QC, Canada

Cristian E. Gutiérrez Department of Mathematics, Temple University, Philadelphia, PA, USA

Annamaria Montanari Dipartimento di Matematica, Università di Bologna, Bologna, Italy

L^∞-Extremal Mappings in AMLE and Teichmüller Theory

Luca Capogna

Abstract These lecture focus on two vector-valued extremal problems which have a common feature in that the corresponding energy functionals involve L^∞ norm of an energy density rather than the more familiar L^p norms. Specifically, we will address (a) the problem of extremal quasiconformal mappings and (b) the problem of absolutely minimizing Lipschitz extensions.

1 Introduction

These notes originate from a C. I. M. E. mini course held by the author in July 2012 in Cetraro, Italy. They are meant to provide a quick introduction to two model L^∞ variational problems involving mappings, i.e. where the set of competitors is not scalar but vector-valued.

The first concerns a classical problem in geometric function theory that first arose in 1928 in the work of Grotzsch [32].

Problem 1. Among all orientation preserving quasiconformal homeomorphisms $w : \Omega \to \Omega'$ whose traces agree with a given mapping $u_0 : \partial\Omega \to \partial\Omega'$, find one which minimizes the functional

$$u \to \left\| \frac{|du|}{(\det du)^{\frac{1}{n}}} \right\|_\infty .$$

L. Capogna (✉)
Department of Mathematical Sciences, Stratton Hall, Worcester Polytechnic Institute 100 Institute Road Worcester, MA 01609, USA
e-mail: lcapogna@wpi.edu

L. Capogna et al., *Fully Nonlinear PDEs in Real and Complex Geometry and Optics*,
Lecture Notes in Mathematics 2087, DOI 10.1007/978-3-319-00942-1_1,
© Springer International Publishing Switzerland 2014

Variants of this problem occur when, instead of using boundary data, the class of competitors is defined in terms of a fixed homotopy class or by requesting that the traces map quasi-symmetrically boundary into boundary.

The second problem has also a classical flavor. It goes back to the work of Whitney [69] and the work of MacShane [49] in 1934, and leads to the recent theory of *absolutely minimal Lipschitz extensions* (AMLE) [5].

Problem 2. Let $\Omega \subset \mathbb{R}^n$, $F \subset \bar{\Omega}$ be a compact set and let $g \in Lip(F, \mathbb{R}^m)$. Among all Lipschitz extensions of F to Ω is there a canonical unique extension that in some sense has the smallest possible Lipschitz norm?

Some natural questions arise in connection to these problems

- Do minimizers exist?
- Are minimizers unique?
- What is the structure of the minimizers? In which norm there is continuity with respect to the data?

The last few decades have seen intense activity from different communities of mathematicians in the study of both problems. However at this time there does not seem to be much synergy and communication between these communities, both in terms of shared techniques used in the study of these problems and in terms of common point of views. One of the goals of these notes is to foster such synergies by outlining some of the common features in these problems. The notes (as well as the lectures) are mainly addressed to graduate students and because of this we have included some very basic material and maintained throughout an informal style of exposition. Since there are no original results in this survey, all proofs are merely sketched, and references to the detailed arguments are provided.

There are several other sources that discuss more extensively either extremal quasiconformal mappings or vector valued AMLE, but we are not aware of a reference striving for a unified point of view. The (possibly too optimistic) goal of this set of notes is to provide such perspective. Regarding other pertinent references: For classical extremal quasiconformal mappings we recommend the surveys of Strebel [64, 65]. Two very clear and extremely well-written accounts of the classical Teichmüller theory can be found in [1, 14]. The paper of Grötzsch [32] is at the origin of the subject and Hamilton's dissertation [33] provided an interesting development. The reader will also benefit from reading the classic monograph [3] as well as the more recent [8]. For the higher dimensional theory of quasiconformal mappings and the corresponding extremal problems I recommend the following fundamental contributions by Gehring and Vaisala [29, 30, 67], as well as the more recent comprehensive book by Iwaniec and Martin [39]. Various aspects of the extremal problem can be found in [6, 7, 9, 23, 60–62]. There is considerably less literature on the vector valued extremal Lipschitz extension problem: A good introduction is in the papers of Barron et al. [11, 12]. More recent developments can be found in the work of Naor and Sheffield [52], Sheffield and Smart [63], Katzourakis [47], Ou et al. [53]. We also want to point out two relevant references that, in our opinion, have great potential for applications to the problems discussed here: Dacorogna and Gangbo [20] and Evans et al. [21].

Although these notes do not involve specific applications, the topic of L^∞ variational problems arises naturally in mathematical models of several real-world phenomena. To this regard, we conclude this introduction with a quote from Robert Jensen's seminal paper [44]

> The importance of variational problems in L^∞ is due to their frequent appearance in applications. The following examples give just a small sample of these. In the engineering of a load-bearing column it is preferable to minimize the maximal stress (i.e., the L^∞ norm of the stress) in the column rather than some average of the stress. When constructing a rocket, the maximal acceleration applied to the payload is an important factor in the design. Optimal operation of a heating-cooling system for an office building requires control of the maximal and minimal temperature within the building rather than the average temperature. Windows on airplanes are made without corners to prevent high pointwise stress concentrations. These considerations motivate the study of the issues of existence, uniqueness, and regularity etc. etc.

2 Notation and Preliminaries

In this section we set the notation for the rest of the notes and include some basic, elementary definitions and results that will be needed later on.

2.1 Notation: Topology

- An *homeomorphism* between two topological spaces is a continuous bijection whose inverse is also continuous.
- A *topological manifold* of dimension $n \in \mathbb{N}$ is a topological space for which every point has a neighborhood homeomorphic to $\mathbb{R}^n$.
- A *smooth manifold* of dimension n is an n-dimensional topological space along with a collection of charts $(U_\alpha, f_\alpha)_{\alpha \in A}$ with $U_\alpha \subset M$ open and such that they cover M, $f_\alpha : U_\alpha \to f_\alpha(U_\alpha) \subset \mathbb{R}^n$ homeomorphism and such that $f_\alpha \circ f_\beta^{-1}$ is smooth on its domain.
- An *homotopy* between two continuous functions f, g between two topological spaces X and Y is a continuous function $H : X \times [0, 1] \to Y$ such that $H(x, 0) = f(x)$ and $H(x, 1) = g(x)$ for all $x \in X$.
- The *fundamental group* $\pi_1(M, p)$ of a topological manifold M with $p \in M$ is the quotient of the space of loops at p through the equivalence relation $\gamma \approx \eta$ if and only if $\gamma \circ \eta^{-1}$ is homotopic to the identity. If $\pi_1(M, p) = 0$ then M is simply connected.
- A *triangle* T in a surface S is a closed set obtained as the homeomorphic image of a planar triangle. The image of vertices and edges of the planar triangle are also called vertices and edges. A *triangulation* of a compact surface S is a finite

set of triangles $T_1, \ldots, T_m$ such that $\cup_{i=1}^m T_i = S$ and every pair T_i, T_j is either disjoint or intersects at a single point (vertex) or a shared edge.

- The *Euler characteristic* of a triangulated compact surface S is given by $\chi = v - e + f$ where v is the number of vertices of the triangulation, e is the number of edges and f the number of triangles. This number does not depend on the specific triangulation of S. The *genus* of S is the number g obtained from the identity $\chi = 2 - 2g$.

Example 1. The sphere has genus zero, as does the unit disc. The torus has genus 1. Roughly, for general orientable surfaces, the genus is the number of *handles* in the surface.

2.2   Notation: Differentials and Dilation of Mappings

The background for fine properties of mappings, their dilation and much more can be found in the monograph [39]. Let $\Omega \subset \mathbb{R}^n$ and denote by

$$u = (u^1, \ldots, u^n) : \Omega \to \mathbb{R}^n \tag{1}$$

a $W_{loc}^{1,n}(\Omega)$ orientation-preserving homeomorphism. At points $x \in \Omega$ of differentiability of u we denote by $du(x) : \mathbb{R}^n \to \mathbb{R}^n$ the *differential of u*. In coordinates one has that for $v \in \mathbb{R}^n$ the action of the differential is[1] $[du(x)(v)]^i = du_{ij}v_j$, $i = 1, \ldots, n$ where we have let $(du)_{ij} = \partial_{x_j} u^i$. Set $|du|^2 = trace(du^T du) = du_{ij} du_{ij}$. At points of differentiability, the pull-back $du^*(x)g_E$ of the Euclidean metric g_E is given by $d(u^*(x)g_E)_{ij} = [du^T du]_{ij} = \partial_{x_i} u^k \partial_{x_j} u^k$, for $i, j = 1, \ldots, n$.

If $n = 2$ it is convenient to use complex notation: Set $u = u^1 + iu^2$, and

$$\partial_z u = \frac{1}{2}(\partial_x u - i\partial_y u) \text{ and } \partial_{\bar{z}} u = \frac{1}{2}(\partial_x u + i\partial_y u).$$

Note $\partial_{\bar{z}} u = \overline{\partial_z \bar{u}}$. We also let $dz = dx + idy$ and note that $dz(\partial_z) = 1$ while $dz(\partial_{\bar{z}}) = 0$. Similarly $d\bar{z} = dx - idy$ and $d\bar{z}(\partial_z) = 0$ and $d\bar{z}(\partial_{\bar{z}}) = 1$.

Next we introduce different ways in which one can quantify how a differentiable homeomorphism $u : \Omega \to \mathbb{R}^n$ can distort the ambient geometry. We start by considering linear bijections $A : \mathbb{R}^n \to \mathbb{R}^n$ expressed in coordinates as $y^i = A_{ij}x^j$ for $i = 1, \ldots, n$. Denote by $|A|_O := \max_{|V|=1} |AV|$ the *operator norm* of A and consider the following quantities

- the *linear dilation* of A is

$$H(A) = \frac{\max_{|h|=1} |Ah|}{\min_{|h|=1} |Ah|}. \tag{2}$$

---

[1]Implicit summation on repeated indices is used throughout the paper.

- the *outer dilation* of A is

$$H_o(A) = \frac{|A|_o^n}{|det\ A|}.$$ (3)

- the *trace dilation* of A is

$$\mathbb{K}(A)^n = \frac{|\sum_{ij} A_{ij}^2|^{n/2}}{|det\ A|}.$$ (4)

If u is as in (1) then we set $\mathbb{K}_u(x) = \mathbb{K}(du(x))$.

2.3 Notation: Complex Analysis

Basic references for the complex analysis background are the classical book of Ahlfors [2] and Jost's monograph [45].

- A C^1 function $w = u + iv : \mathbb{C} \to \mathbb{C}$ is *holomorphic* if

$$\partial_{\bar{z}} w = u_x + iu_y + iv_x - v_y = 0$$

  Equivalently w must satisfy the *Cauchy-Riemann equations* $u_x = v_y$ and $u_y = -v_x$.
- *Conformal invariance of harmonic functions.* If $w = h(z)$ is a holomorphic function and $f : \mathbb{C} \to \mathbb{C}$ is smooth then

$$\frac{\partial^2}{\partial z \partial \bar{z}} f \circ h(z) = \frac{\partial}{\partial z} \left[\frac{\partial f}{\partial w} \frac{\partial h}{\partial z} + \frac{\partial f}{\partial \bar{w}} \frac{\partial \bar{h}}{\partial z} \right]$$

$$= \frac{\partial}{\partial z} \left[\frac{\partial f}{\partial \bar{w}} |_{h(z)} \frac{\partial \bar{h}}{\partial z} \right] = \left[\frac{\partial^2}{\partial w \partial \bar{w}} f \right] |_{h(z)} \partial_z h \partial_{\bar{z}} \bar{h}.$$

- A holomorphic map $u : U \subset \mathbb{C} \to \mathbb{C}$ is called *conformal* if $\partial_z u \neq 0$ at every point in U.

Example 2. Set

$$D = \{z \in \mathbb{C} |\ |z| < 1\} \text{ and } H = \{z = x + \mathbf{i}y|\ y > 0\}$$

These are conformally equivalent under the map $H \to D$ given by $z \to \frac{z-z_0}{z-\bar{z}_0}$.

Theorem 1. *Every* $f : D \to D$ *(or* $f : H \to H$*) which is biholomorphic (i.e., conformal and bijective) is a Möbius transformation, i.e. there are* $a, b, c, d \in \mathbb{C}$ *such that*

$$f(z) = \frac{az + b}{cz + d}.$$

For any ring R define the group

$$SL(2, R) = \left\{ \begin{pmatrix} a & b \\ c & d \end{pmatrix} \middle| ad - cb = 1 \right\}$$

while $PSL(2, R)$ denotes its quotient by the sub-group generated by $\pm Id$. Every element in $PSL(2, R)$ defines a Möbius transformation $H \to H$.

Definition 1. A group G acts as a transformation group on a manifold M if there is a map $G \times M \to M$ denoted as $(g, x) \to gx$ with $(g_1 g_2)(x) = g_1(g_2 x)$ and $ex = x$. The isotropy group of $x \in M$ is a subgroup of G which fixes x.

Example 3. The group $PSL(2, \mathbb{R})$ acts as a transformation group of H. The isotropy group of each element is isomorphic to $SO(2)$.

Both D and H can be given a (non-Euclidean) metric structure through the *hyperbolic metric*

$$\frac{1}{y^2} dz d\bar{z} \text{ on } H \quad \text{and} \quad \frac{1}{(1 - |z|^2)^2} dz d\bar{z} \text{ on } D.$$

An isometry between two Riemannian manifolds

$$u : (M, g) \to (M', g')$$

is a map such that

$$g'_{u(x)}(d_x u V, d_x u W) = g_x(V, W)$$

for any $x \in M$ and $V, W \in T_x M$.

Theorem 2. *All isometries between the hyperbolic H and D are Möbius transformations. The isometry group of H is $PSL(2, \mathbb{R})$.*

Definition 2. A group action G on M is *properly discontinuous* if every $x \in M$ has a neighborhood U such that $\{g \in G | gU \cap U \neq 0\}$ is finite and if x, y are not in the same orbit then they have neighborhoods U_x, U_y such that $gU_x \cap U_y = \emptyset$ for all $g \in G$.

Definition 3. Let $\Gamma \subset PSL(2, \mathbb{R})$ be properly discontinuous subgroup and $z_1, z_2 \in H$. We say that z_1 and z_2 are equivalent if there exists $g \in \Gamma$ such that $gz_1 = z_2$. Consider H/Γ the space of quotient classes equipped with the quotient topology.

Proposition 1. *Let $\Gamma \subset PSL(2, \mathbb{R})$. If the action of Γ on H properly discontinuous and does not fix points ($gx \neq x$ for all $x \in H$ and all $g \neq id$) then the quotient H/Γ can be given a Riemann surface structure.*

3 Conformal Deformations

An a.e. differentiable homeomorphism $u : \mathbb{R}^n \to \mathbb{R}^n$ is *conformal* if there exists a scalar function λ such that at every point of differentiability

$$du^T du = \lambda Id \qquad (5)$$

The pull-back of the Euclidean metric dx^2 is a scalar multiple of dx^2, i.e. *angles are preserved*. Equivalently, a.e. in Ω, one must have

$$g := \frac{du^T du}{(\det du)^{\frac{2}{n}}} = Id. \qquad (6)$$

The function $\sqrt{trace(g)} = |du|/(\det du)^{1/n}$ is called *dilation* of u.

Remark 1. At every point of differentiability for u one has $\mathbb{K}_u = trace(g) \geq n$, with the equality being achieved if and only if $g = Id$.

Definition 4. Following Ahlfors (see also [39]) we define the *distortion tensor of u at a point of differentiability $x \in \mathbb{R}^n$*

$$S(g) := \frac{g + g^T}{2} - \frac{trace(g)}{n} Id = g - \frac{trace(g)}{n} Id, \qquad (7)$$

and denote by

$$\mathbb{K}(u, \Omega) = ||\mathbb{K}_u||_{L^\infty(\Omega)} = ||\sqrt{trace(g)}||_{L^\infty(\Omega)}, \qquad (8)$$

the *maximal dilation* of u in Ω.

Proposition 2. *With the notation above, one has that a diffeomorphism u is conformal if and only if $S(g) = 0$ and if and only if $\mathbb{K}(u, \Omega)^2 = \mathbb{K}_u^2 = trace(g) = n$ identically in Ω.*

Remark 2. It is not difficult to show that if $\mathbb{K}_u = K_0 > \sqrt{n}$, then there exists $\varepsilon = \varepsilon(K_0) > 0$ so that

$$\varepsilon \leq |S(g)|^2 \leq \mathbb{K}_u^4 \left(1 - \frac{1}{n}\right).$$

When $n = 2$, if we denote by $0 \leq \lambda_1 \leq \lambda_2$ be the eigenvalues of g, then one can find an explicit lower bound. In this case, $\lambda_1 \lambda_2 = 1$ and

$$|S(g)|^2 = \lambda_1^2 + \lambda_2^2 - \frac{1}{2}(\lambda_1 + \lambda_2)^2 = \frac{1}{2}(\lambda_1 + \lambda_2)^2 - 2\lambda_1\lambda_2 = \frac{1}{2}(\mathbb{K}_u^4 - 4).$$

Remark 3. Denote by $CO_+(n)$ the space of differentials of orientation preserving conformal mappings, then its tangent space $TCO_+(n)$ at the identity is

$$\left\{ A \in \mathbb{R}^{n \times n} s.t.\ S(A) = \frac{A + A^T}{2} - \frac{trace(A)}{n} Id = 0 \right\}.$$

Accordingly we have that the distance of a matrix A from $CO_+(n)$ satisfies

$$d^2(A, CO_+(n)) = c|S(A - I)|^2 + O(|A - I|^4).$$

This shows that the operator S arises naturally when considering the linearization of the distance of a deformation from being conformal. For more results from this point of view, including a remarkable geometric rigidity result in the spirit of Frieseke et al. [24], see the work of Faraco and Zhong [22].

Three remarkable properties of conformal deformations

- *Conformal implies smooth.* If an homeomorphism $u \in W_{loc}^{1,n}(\Omega, \mathbb{R}^n)$ satisfies $\mathbb{K}(u, \Omega) = \sqrt{n}$ then a result of F. Gehring [28] implies that $u \in C^\infty(\Omega)$. The proof is based on regularity of weak solutions to the n-Laplacian, via the De Giorgi-Nash-Moser theorem. See the discussion below on Liouville theorem for more details. Moreover, if f is a weak solution to the n-Laplacian and u is conformal then $f \circ u$ is also n-harmonic. This is the so-called *morphism* property.
- For $n = 2$; *Conformal transformation are holomorphic diffeomorphism and viceversa.* In particular the space of conformal planar deformations is infinite dimensional.
- *Riemann mapping theorem.* Any non-empty, simply connected open planar set can be mapped conformally to the disc (uniquely if one prescribes target for *three* points).

Rigidity of conformal deformations. Despite the flexibility of the Riemann mapping theorem and the usefulness in changes of variables arguments, conformal mappings exhibit aspects of rigidity that make it too restrictive for many applications.

- *Liouville theorem.* For $n \geq 3$ conformal deformations are compositions of translations, rotations, dilations and inversions. The theorem was proved originally by J. Liouville [48] with the hypothesis that the fourth order derivatives of the maps be continuous. Gehring [28] and Reshtnyak [57] established remarkable generalizations respectively to quasiconformal and to quasiregular mappings in $W_{lot}^{1,n}$. For $n = 2l$ a sharp form of the Liouville theorem was established by Iwaniec and Martin in [38]. In this paper, among other things, it is proved that for $l > 1$, every $u \in W_{loc}^{1,l}(\Omega, \mathbb{R}^{2l})$ with $\det du \geq 0$ (or $\det du \leq 0$) a.e. and such that $H(du) = 1$, i.e. $\|du\|_O \leq \min_{|v|=1} |duv|$ a.e. is either constant or the restriction of a Möbius transformation to Ω. The Sobolev exponent l is optimal in the sense that there are weak $W_{loc}^{1,p}$ solutions of the Cauchy-Riemann equations with $p < l$, which are not Möbius.

- *Rigidity with respect to boundary data.* Even in the plane, despite the Riemann mapping theorem one *cannot prescribe boundary data* (more than three points) when mapping conformally one domain into the other. For instance, in mapping one rectangular box into another, sending sides to sides, one can achieve this through a conformal deformation only if the boxes are similar (see the next section).

The intrinsic rigidity of conformal mappings provided a motivation for the extension to a larger family, that of *quasiconformal mappings*. Quoting F. Gehring [29],

> ...quasiconformal mappings constitute a closed class of mappings interpolating between homeomorphisms and diffeomorphisms for which many results of geometric topology hold regardless of dimension.

In the next section we will see how at the genesis of the theory of quasiconformal mappings lies a L^∞ variational problem.

4   Grötzsch Problem and Quasiconformal Deformations

Let R and R' be two rectangles with sides a, b and a', b', that are not similar (i.w. $a/b \neq a'/b'$). It is then easy to see that there is no conformal deformation mapping R to R' sending edges to edges. In connection to this observation, in 1928 H. Grötzsch [32] posed the following question

Problem 3. Is there a *most nearly conformal mapping* between R and R'?

Quoting L. Ahlfors [1] in relation to this problem

> This calls for a measure of approximate conformality, and in supplying such a measure Grötzsch took the first step toward the creation of a theory of q.c. mappings.

To address Grötzsch's question one would need to identify a quantitative way of determining how non-conformal a mapping can be and then find an extremal point for this quantity in a suitable class of competitors. For such a general scheme to work it is of paramount importance to have good compactness properties for the class of competitors. Such considerations hint at the *need of introducing a more general class of deformations that are less rigid, yet retain some of the useful features of conformal mappings*. One also would like to have an instrument to quantify how far a given deformation is from being conformal.

Definition 5. Let $\Omega \subset \mathbb{R}^n$ be an open set. If $u \in W^{1,n}_{loc}(\Omega, \mathbb{R}^n)$ is an homeomorphism then we say it is *quasiconformal* if

$$\mathbb{K}(u, \Omega) = \|\sqrt{trace(g)}\|_{L^\infty} = \left\| \frac{|du|}{(\det du)^{1/n}} \right\|_{L^\infty} < \infty.$$

We say u is K-quasiconformal if $K = \|H_O(du)\|_\infty = \||du|_O/\det du^{1/n}\|_\infty$.

Example 4. In the following we list some simple examples of quasiconformal mappings.

- Linear bijections $x \to Ax$ with $A \in \mathbb{R}^{n \times n}$.
- Diffeomorphisms with non-vanishing Jacobians are locally quasiconformal.
- For $a \neq 0$ consider the family of quasiconformal mappings $u(x) = |x|^{a-1}x$. For $a = -1$ this is the conformal inversion.
- In cylindrical coordinates (r, ϕ, z) set $D_\alpha = \{\phi \in (0, \alpha)\}$ and define $f : D_\alpha \to D_\beta$ as $f(r, \phi, z) = (r, \beta\phi/\alpha, z)$ *(folding map)*.
- In spherical coordinates (R, ϕ, θ) define C_α a cone of angle α by $0 \leq \theta < \alpha$. Set $f : C_\alpha \to C_\beta$ as $f(R, \phi, \theta) = (R, \phi, \beta\theta/\alpha)$. The map is quasiconformal for $\beta < \pi$ and but fails to be quasiconformal for $\beta = \pi$.

Definition 5 seems to require a-priori information on a.e. differentiability of the mapping which are counterintuitive in relation to the need for compactness of the class of competitors we referred to. There are in fact previous equivalent definitions for quasiconformality which do not require *any a priori differentiability*.

Definition 6. *Geometric definition* Let $r > 0$ and $x \in \Omega$. Set

$$L(x, r) = \sup_{y \in \Omega| \ |x-y| \leq r} |u(x) - u(y)|$$

and

$$l(x, r) = \inf_{y \in \Omega| \ |x-y| \geq r} |u(x) - u(y)|$$

The homeomorphism u is quasiconformal if there exists $H \geq 1$ such that for *every* $x \in \Omega$ the *linear dilation* satisfies

$$H(x, u) := \limsup_{r \to 0} \frac{L(x, r)}{l(x, r)} \leq H < \infty. \tag{9}$$

Remark 4. If $A : \mathbb{R}^n \to \mathbb{R}^n$ is a bijection then $H(x, A) = H(A)$ with $H(A)$ defined as in (2). If $u : \mathbb{R}^n \to \mathbb{R}^n$ is differentiable at the point x with non-vanishing Jacobian determinant then $H(x, u) = H(du(x))$.

Remark 5. The differential $du(x)$ transforms circles centered at the origin into similar ellipses. The quantity $H(du)$ is the ratio of the axis of such ellipse. Thus quasiconformal deformations map infinitesimal circles into ellipses with a bounded ratio of the axis.

Homemorphism for which (9) holds with lim sup substituted by sup are called *quasisymmetric*. It was F. Gehring [27] who first proved that quasiconformal implies quasisymmetric if $n \geq 2$. See also [34] for quantitative estimates and extensions to more general metric spaces.

Theorem 3. *Consider an homeomorphism* $u : \Omega \to \Omega'$, *then the quantity* $\|H(x, u)\|_{L^\infty}$ *is finite if and only if* $u \in W^{1,n}_{lot}(\Omega, \mathbb{R}^n)$ *and* $\mathbb{K}(u, \Omega)$ *is finite.*

Theorem 4 (Gehring, 1962). *For every* $K \geq 1$ *and* $n \in \mathbb{N}, n \geq 2$ *there exits* $\theta_{n,K} : (0,1) \to \mathbb{R}$ *increasing, with* $\lim_{r \to 0} \theta_{n,K}(r) = 0$ *and* $\lim_{r \to 1} \theta_{n,K}(r) = \infty$ *such that for every* $f : \Omega \to \Omega'$ K-*quasiconformal one has*

$$\frac{d(f(x), f(y))}{d(f(x), \partial\Omega')} \leq \theta_{n,K}\left(\frac{d(x,y)}{d(x, \partial\Omega)}\right),$$

for all distinct $x, y \in \Omega$ *sufficiently close. Moreover for* r *sufficiently small, one can choose* $\theta_{n,K}(r) = c_n r^\alpha$ *with* $\alpha = K^{1/(1-n)}$.

In complex notation one denotes the map as

$$z = x + \mathbf{i}y \in \Omega \subset \mathbb{C} \to \zeta(z) = \xi + i\eta,$$

and set $p = \partial_z \zeta$ and $q = \partial_{\bar{z}} \zeta$, so that $d\zeta = pdz + qd\bar{z}$. The mapping $d\zeta$ is affine and satisfies

$$\left|\,|p| - |q|\,\right| |dz| \leq |d\zeta| \leq \left|\,|p| - |q|\,\right| |dz|$$

From the latter we see that the ratio of the axes of the ellipse obtained as image of a circle under $d\zeta$ is given by the maximal dilation

$$K = \sup_\Omega \frac{|p| + |q|}{|p| - |q|}.$$

We also define the maximal excentricity $\kappa = \frac{K-1}{K+1} = \sup \frac{|q|}{|p|}$. Note that ζ is conformal iff $K = 1, \kappa = 0$. The Jacobian determinant of the map ζ is $J = |p|^2 - |q|^2$.

Remark 6. Since the derivatives of the inverse map $\zeta \to z(\zeta)$ are given by

$$p' = J^{-1}\bar{p} \text{ and } q' = -J^{-1}q$$

then the mapping $\zeta = \zeta(z)$ and $z = z(\zeta)$ have the same dilation at corresponding points, hence the same maximal dilation. Moreover, the dilation is invariant by conformal deformation in both the z and the ζ planes.

4.1 Grötzsch Problem

Let us return to Grötzsch original question. Consider two rectangles R, R' with sides parallel to the axis and with one vertex at the origin, as illustrated in Fig. 1.

Remark 7. The affine transformation mapping $R \to R'$ that maps edges to edges is the *anisotropic dilation* $\xi = \frac{a'}{a}x$ and $\eta = \frac{b'}{b}y$ i.e.,

$$\zeta = \frac{1}{2}\left([\frac{a'}{a} + \frac{b'}{b}]z + [\frac{a'}{a} - \frac{b'}{b}]\bar{z}\right). \tag{10}$$

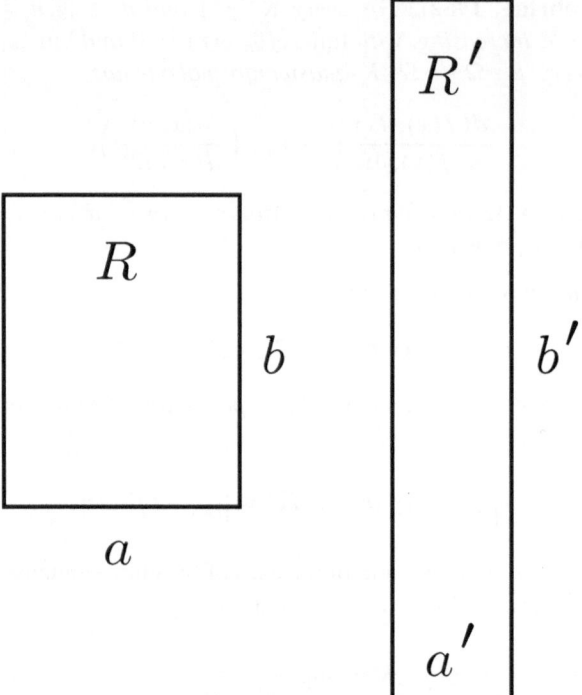

Fig. 1 Grötzsch problem

The dilation of the affine map is a constant

$$
K = \begin{cases} \frac{a'b}{b'a} & \text{if} \quad \frac{a'}{b'} > \frac{a}{b} \\ \frac{ab'}{ba'} & \text{if} \quad \frac{a'}{b'} < \frac{a}{b} \end{cases} \tag{11}
$$

Proposition 3. *Every diffeomorphism from* R *to* R', *mapping edges to edges, has dilation larger or equal than the dilation of the affine transformation* (10). *Moreover, if its dilation is the same as that in* (11) *then the diffeomorphism must coincide with* (10).

Proof. Let $\zeta : R \rightarrow R'$ be diffeomorphism from R to R', mapping edges to edges. Recall that for $\xi(0, y) = 0$ while $\xi(a, y) = a'$. A simple computation shows that the integral of the differential form $\xi(x, y)dy$ along the boundary of R yields $a'b = \int_{\partial R} \xi(x, y)dy$. On the other hand Stokes theorem yields

$$
a'b = \int_{\partial R} \xi(x, y)dy = \int_{R} \partial_x \xi(x, y)dxdy = \int_{R} Re(p + q)dxdy
$$

(since $(|p| + |q|)^2 \leq K(|p|^2 - |q|^2)$)

$$\leq \int_R |p| + |q| dxdy \leq \sqrt{|R|} \left(\int_R (|p| + |q|)^2 dxdy \right)^{1/2} \leq \sqrt{K|R||R|'}.$$

In conclusion $\frac{a'b}{ab'} \leq K$. Reverting the role of z and ζ and recalling that the two maps have the same dilation one has $\frac{ab'}{a'b} \leq K$. *In both cases the affine map has minimal dilation.*

To prove the second part of the proposition we notice that the only way one may have equality in the previous computation is if

$$Re(p + q) = |p| + |q| \text{ and } (|p| + |q|)/(|p| - |q|)$$

is constant. The former yields $Im(p)=Im(q)=0$ and consequently $\partial_y \xi = \partial_x \eta = 0$. The latter yields that $\partial_x \xi(x) = K \partial_y \eta(y)$ which has as immediate consequence that $\partial_x \xi(x) = const$ and $\partial_y \eta(y) = const$. Hence any extremal map must have the form $\zeta(z) = \alpha x + \mathbf{i}\beta y$ and α, β must match the values for the affine map for ζ to map R into R', vertex by vertex. □

4.2 Grötzsch Problem Revisited

Consider two Jordan regions $Q, Q' \subset \mathbb{C}$ with distinguished boundary points $p_1, \ldots, p_4 \in \partial Q$ and $p'_1, \ldots, p'_4 \in \partial Q'$.

Problem 4. Among all diffeomorphism $\zeta : Q \to Q'$ mapping $\zeta(p_i) = p'_i$, $i = 1, \ldots, 4$, find the one with minimal maximal dilation.

As illustrated in Fig. 2, the Riemann mapping theorem yields two rectangles R, R' and conformal transformations $\phi : Q \to R$ and $\psi : Q' \to R'$ of the domains Q, Q' to R, R' with the points p_i, p'_i mapped to the vertices of the rectangles. Since the dilation is conformally invariant any map $\zeta : R \to R'$ has the same dilation of its lift $\phi \circ \zeta \circ \psi^{-1} : Q \to Q'$. The previous argument yields the following conclusion: *The extremal map for the revisited Grötzsch problem is a composition of an affine anisotropic dilation with conformal transformations $\phi \circ affine \circ \psi^{-1}$. Such map is unique modulo conjugation with conformal transformations.*

Such transformations are examples of *Teichmüller mappings*.

5 Teichmüller Theorem and Extremal Quasiconformal Mappings

Let $u : \Omega \to \mathbb{R}^n$ be a quasiconformal mapping. Since $\mathbb{K}(u, \Omega) = \sqrt{n}$ if and only if the mapping u is conformal, we will interpret Grötzsch's *closest to conformal* requirement as *minimizing* $\mathbb{K}(u, \Omega) = ||\sqrt{trace(g)}||_{L^\infty}$ among all competitors. The

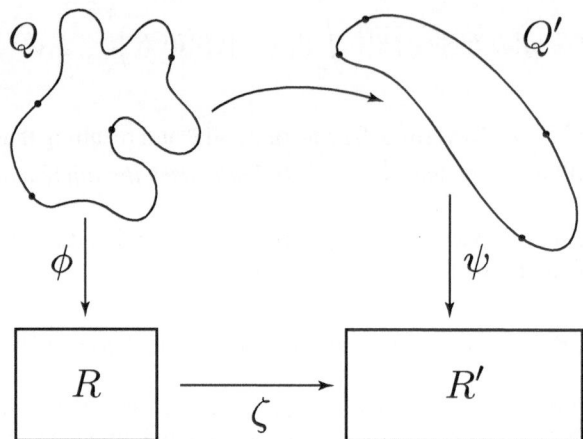

Fig. 2 Grötzsch problem revisited: General Jordan regions

$n = 2$ setting has been studied in depth by many mathematicians. Here we recall the work of Reich [54], Reich and Strebel [55, 56], Strebel [65] and Gardiner [26].

In the following we will briefly highlight the set of ideas, techniques and results loosely known as *Teichmüller theory*. This important theory has as a starting point a similar L^∞ variational problem in the context of holomorphisms between Riemann surfaces of same genus ($g > 3$) and where the constraint defining the class of competitors is given by membership in the same homotopy class. In this setting one has existence, uniqueness and some amount of regularity for the minimizers.

For $n > 2$ less is known. The fundamental reference by Gehring and Vaisala [30] establishes the problem in a more general setting and provides some existence results. The higher dimensional analogue of Grötzsch problem was solved by Fehlmann [23]. A great amount of recent literature focuses on the L^p variational problems, which we will briefly describe through the work of Astala et al. [7]. We also recall related work of Balogh et al. [10] and Astala et al. [9].

In a (rough) comparison with similar problems in elasticity, *conformal deformations correspond to isometries.* Accordingly, the variational problems stated above corresponds to finding deformations *closest to isometries* in given classes of competitors.

Some of the main obstacles in studying this L^∞ problem are

- Lack of convexity.
- L^∞ functionals are not sensitive to deformations of functions away from their maximum. Unlike L^p averages they are not "local". *This makes uniqueness unlikely.*
- The problem is vector-valued, and as such not approachable through the established techniques from game theory or viscosity solutions.
- There is a topological constraint.

5.1 Riemann Surfaces

A *Conformal Atlas* on a two dimensional smooth manifold is an atlas $(U_\alpha, z_\alpha)_{\alpha \in A}$ with $z_\alpha : U_\alpha \to \mathbb{C}$ local charts such that the transition maps $z_\beta \circ z_\alpha^{-1} : z_\alpha(U_\alpha \cap U_\beta) \to z_\beta(U_\alpha \cap U_\beta)$ are holomorphic.

An atlas (U_α, z_α) is compatible to another atlas (V_β, w_β) if their union is still a conformal atlas. A *conformal structure* is the union of an atlas with all other compatible charts.

Definition 7. A *Riemann surface* is a two dimensional smooth manifold with a conformal structure.

Example 5. The *Riemann sphere* $S^2 = \{x_1^2 + x_2^2 + x_3^2 = 1\} \subset \mathbb{R}^3$. To show that S^2 is a Riemann surface we consider an atlas with open sets U_1, U_2 obtained from the whole sphere minus respectively the north pole $(0, 0, 1)$ and the south pole $(0, 0, -1)$. Define the *stereographic projections* charts

$$z_1(x) = \frac{x_1 + ix_2}{1 - x_3} \text{ on } U_1 \text{ and } z_2(x) = \frac{x_1 - ix_2}{1 + x_3} \text{ on } U_2$$

Note that $z_1(U_1 \cap U_2) = z_2(U_1 \cap U_2) = \mathbb{C} \setminus \{0\}$ and that

$$z_2 \circ z_1^{-1}(a + ib) = \frac{a + ib}{a^2 + b^2} = \frac{1}{a + ib}$$

Example 6. The *Riemann Torus* is defined as follows: Set $w_1, w_2 \in \mathbb{C}$ two non-zero vectors and define an equivalence relation in $\mathbb{C}$ by saying that $a + ib \approx x + iy$ if there exists two rational numbers m, n such that $a + ib = x + iy + mw_1 + nw_2$. The discrete Abelian subgroup $M = \{mw_1 + nw_2\}$ is called a *lattice*. If we let π be the projection to the quotient space then $\mathbb{T} = \pi(\mathbb{C})$ is a Riemann surface. To define an atlas we consider open sets $O \subset \mathbb{C}$ containing to equivalent pairs (for instance a subset of a fundamental domain) and define the chart $U = \pi(O)$ and $z = \pi|_O^{-1}$. Since $z_\alpha \circ z_\beta^{-1}$ is a translation then this is a conformal atlas.

We have already seen that if $\Gamma \subset PSL(2, \mathbb{R})$ and the action of Γ on H is properly discontinuous and does not fix points then the quotient H/Γ can be given a Riemann surface structure. Viceversa one has the following

Theorem 5 (Uniformization theorem). *Let Σ be a compact Riemann surface of genus p. There exists a conformal diffeomorphism $f : \Sigma \to \Sigma'$ where Σ' is either*

(i) *of the form H/Γ if $p \geq 2$;*
(ii) *A torus $\mathbb{C}/M$ if $p = 1$;*
(iii) *the Riemann sphere if $p = 0$.*

As corollary, the universal cover of a compact Riemann surface is conformally equivalent to S^2, $\mathbb{C}$ or D.

Definition 8. A continuous map $u : S \rightarrow S'$ is holomorphic if it is so when expressed (locally) through conformal charts. If these local expression have non vanishing $\partial_z u$ derivative then u is conformal.

Let us recall the topological classification of compact Riemann surfaces

Theorem 6. *Two differentiable, orientable, compact triangulated surfaces[2] are homeomorphic if and only if they have the same genus.*

Every Riemannian metric $g_{ij}dx_i dx_j$ on a oriented surface can be written locally in complex coordinates as

$$\sigma(z)|dz + \mu(z)d\bar{z}|^2 = \sigma(z)(dz + \mu d\bar{z})(\bar{z} + \bar{\mu}dz)$$

where $\sigma > 0$ (real) and $|\mu| < 1$.

Theorem 7. *Every oriented Riemannian surface admits a conformal structure and a conformal Riemann metric $\lambda dz d\bar{z}$. A local system of holomorphic coordinates is given by the solutions of the equation $\partial_{\bar{z}} u = \mu \partial_z u$ and $\lambda = \frac{\partial \sigma}{\partial_z u \partial_{\bar{z}} \bar{u}}$.*

5.2 Teichmüller Theorem

The focus of Teichmüller theorem is on a classification of all possible conformal structures of a given Riemann surface S, and on establishing a structure theorem for such a moduli space. The natural candidate for *space of all conformal structures* is

Definition 9. Given a compact Riemann surface S with genus p, we define the *moduli space* M_p of conformal structures on S where (S, g_1) and (S, g_2) are identified if there exists a conformal diffeomorphism between them.

However this moduli space does not have a manifold structure and its topology is very complicated. To somewhat simplify the structure Teichmüller proposed a new notion of moduli space of conformal structures, known today as Teichmüller space.

Definition 10. Given a compact Riemann surface S with genus p, we define the *moduli space* T_p of conformal structures on S where (S, g_1) and (S, g_2) are identified if there exists a conformal diffeomorphism between them which is homotopic to the identity.

The first step in studying the structure of this space is given by the following existence theorem

---

[2]Recall that every Riemann surface is orientable and any conformal atlas yields a triangulation.

Theorem 8 (Existence Theorem). *Given S, S' closed Riemann surfaces of same genus and $\alpha : S \to S'$ an homeomorphism, there exists a quasiconformal mapping $\zeta : S \to S'$ homotopic to α and minimizing the maximal dilation in the homotopy class of α.*

What is needed next is a uniqueness result for such minimizers as well as an algebraic characterization that would allow to define a manifold structure on the moduli space. In the next sections we will state such uniqueness results and then proceed to sketch Ahlfors' proof of this remarkable characterization.

5.3   Coverings and Group Action

If S, S' are Riemann surfaces of same genus $g > 1$ realized as $D/G, D'/G'$. The quotient map $p : D \to D/G = S$ is a covering map and the group G, which acts on D is called a Fuchsian group. The disc D is the universal cover of S.

Theorem 9. *Any homeomorphism map $\zeta : S \to S'$ can be lifted to a family of mappings $\zeta : D \to D'$ with the property that for every $g \in G$ there exists a unique $g' := \alpha(g) \in G'$ such that*

$$\zeta(g(z)) = g'(\zeta(z)).$$

Viceversa, any homeomorphism $\zeta : D \to D'$ which satisfies the identity above induces an homeomorphism $\zeta : S \to S'$. Lifts of quasiconformal mappings are quasiconformal with the same dilation. The maps $g \to \alpha(g)$ are group isomorphisms. Any two lifts are related by a inner automorphism of G or G'.

Theorem 10. *Any two homeomorphism maps $S \to S'$ are homotopic if and only if the determine isomorphisms $G \to G'$ which differ only by an inner automorphism. So essentially, modulo renormalization, there exists a one-to-one correspondence between homotopy classes of homeomorphisms and isomorphisms $G \to G'$.*

This result allows to reframe Teichmüller theorem and the variational problem only in terms of mappings $\zeta : D \to D'$ which satisfy the functional equation $\zeta(g(z)) = g'(\zeta(z))$.

Theorem 11 (Existence Theorem reformulated). *Given $\zeta_0 : D \to D'$ a fixed homeomorphism, let $\alpha : G \to G'$ denote the induced isomorphism. There exists a quasiconformal mapping $\zeta : D \to D'$ minimizing the maximal dilation in the class of all homeomorphisms satisfying the function equation*

$$\zeta(g(z)) = \alpha(g)(\zeta(z)).$$

Consider the set of all quasiconformal mappings satisfying the identity (this is not empty since the surfaces are diffeomorphic) and with dilation less than a fixed number K. Gehring's theorem implies that this is a normal family, and hence any minimizing sequence will converge to either a quasiconformal mapping satisfying the same functional identity or a constant. Constant are ruled out by the functional identity and the fact that no element of D' is fixed by every $g' \in G'$. Uniqueness of the representative is provided by a deep connection between extremal quasiconformal mappings and *quadratic differentials*.

5.4   Quadratic Differentials

Consider a 1-form $dz = dx + \mathbf{i}dy$ in $\mathbb{C}$. If $F : \mathbb{C} \to \mathbb{C}$ is a conformal map and we denote $z(w) = F(w)$ then for any $(a, b) \in \mathbb{R}^2$ we can compute the action of $dF : \mathbb{R}^2 \to \mathbb{R}^2$ in complex notation as $dF \cdot (a, b) = \frac{dF}{dw}(a + ib)$. In view of this then the pull-back F^*dz is given by $F'(w)dw = \frac{dz}{dw}dw$, in fact the action of F^*dz on any complex tangent vector $a + ib$ can be computed through

$$F^*dz(a + ib) = dz(dF(a + ib)) = dz(F'(w)(a + ib)) = F'(w)dw(a + ib)$$

Through a similar computation one sees that the symmetric 2-tensor $\phi(z)dz^2$ pulls back to $\phi(z(w))(\frac{dz}{dw})^2 dw^2$.

These computations motivate the following

Definition 11. Let S be a Riemann surface and $\{(U_\alpha, h_\alpha)\}$ denote its conformal structure. A meromorphic (resp. holomorphic) *quadratic differential h* in S is a set of meromorphic (res. holomorphic) functions f_α in the local coordinates given by $z_\alpha = h_\alpha(p)$ with $p \in S$ satisfying the transformation law

$$f_\alpha(z_\alpha) = f_\beta(z_\beta)\left(\frac{\partial z_\beta}{\partial z_\alpha}\right)^2,$$

for all charts (U_α, z_α) and (U_β, z_β) around a point $p \in S$.

Recalling the formula for the pull back of a complex 2-tensor described earlier we can write the definition above as

$$h_\alpha(z_\alpha)dz_\alpha^2 = h_\beta(z_\beta)dz_\beta^2$$

Observe that quadratic differentials are holomorphic sections of the bundle of holomorphic symmetric tensors.

Let $Q(S)$ denote the space of all quadratic differentials on a given compact Riemann surface S. Since the sum of two quadratic differentials as well as the multiplication by a scalar of a quadratic differential are still elements of $Q(S)$ then the latter is a complex vector space. The following is a consequence of the Riemann-Roch Theorem

Theorem 12. *The space $Q(S)$ has finite dimension (over $\mathbb{R}$) $6p - 6$.*

Theorem 13 (Structure of minimizers). *Given S, S' closed Riemann surfaces of same genus $p > 1$ and $\alpha : S \to S'$ an homeomorphism. Denote by $\zeta : S \to S'$ a quasiconformal mapping homotopic to α and minimizing the maximal dilation in the homotopy class of α. Either ζ is analytic or there exists a quadratic differential fdz^2 on S and a constant $\kappa \in (0, 1)$, such that ζ is differentiable away from the zero set of f (with non-vanishing complex derivatives q, p) and satisfies*

$$\frac{q}{p} = \kappa \frac{\bar{f}}{|f|}.$$

The quadratic differential is uniquely represented up to a positive constant factor and κ represents the (constant) eccentricity of the extremal mapping.

Theorem 14 (Uniqueness). *Every map ζ whose complex derivatives satisfy*

$$\frac{q}{p} = \kappa \frac{\bar{f}}{|f|}$$

has a maximal dilation which is strictly smaller than the dilation of any other mapping (not conformally equivalent to ζ).

These result yield that for every homotopy class one has existence of a unique minimizer for the maximal dilation and associated to this minimizer there is a unique pair of quadratic differentials. This correspondence gives a manifold structure to the Teichmüller space, with the same dimension $6p - 6$ as the space of quadratic differentials.

5.5 Ahlfors' Proof of Existence and Uniqueness

In [1], Ahlfors considers the following m-mean distortion functional: For every a. e. differentiable map $\zeta = \xi + i\eta : D \to D'$ and $m \geq 1$,

$$I_m(\zeta) = \frac{1}{\pi} \int \int_{D'} \left(\frac{|p|^2 + |q|^2}{|p|^2 - |q|^2} \right)^m \Bigg|_{\zeta^{-1}(\xi + i\eta)} d\xi d\eta.$$

The customary 1-parameter deformations used in the calculus of variations to derive the Euler-Lagrange equations are of the form $\zeta_s = \zeta + s\psi$ where $s \in (-\varepsilon, \varepsilon)$ and $\phi \in C_0^\infty(D, D')$ serves as a test function. However if ζ is merely quasiconformal, in particular not a C^1 diffeomorphism then the deformation $\zeta_s = \zeta + s\psi$ may fail to be an homeomorphism and hence be outside of the set of competitors, making it useless for the purpose of deriving a PDE which describes the behavior of minimizers.

To circumvent this problem one may choose to do a different set of perturbations, acting on the domain of the map, rather than one the image, thus setting:

$$z = H(z', \varepsilon) := s' + \varepsilon h(z') + o(\varepsilon),$$

yielding

$$\partial_z H = 1 + \varepsilon \partial_z h + o(\varepsilon); \text{ and } \partial_{\bar{z}} H = \varepsilon \partial_{\bar{z}} h + o(\varepsilon).$$

If G is a Fuchsian group acting on a Riemann surface S then for H to determine a deformation of the surface one needs $H(gz, \varepsilon) = gH(z, \varepsilon)$ for every $g \in G$. This eventually yields $h(gz) = \partial_z g \, h(z)$, which characterizes all *infinitesimal deformations of S*.

Remark 8. A brief digression: If one considers the Dirichlet energy

$$\int \int_D |\partial_z w|^2 + |\partial_{\bar{z}} w|^2 dx dy$$

then the usual *exterior* deformations lead to the Laplacian $\partial_z \partial_{\bar{z}} w = 0$.

If instead one proceeds as in Ahlfors (and Hopf, Morrey, and many others) and carries out *inner variations* as described earlier then one obtains the PDE

$$\partial_{\bar{z}} \left(\partial_z w \, \overline{\partial_{\bar{z}} w} \right) = 0,$$

which is of a very different nature from Laplace's equation. To the best of our knowledge, currently the sharpest regularity result known for such PDE is Lipschitz continuity, see Iwaniec et al. [43]. See also earlier work of Bauman et al. [13].

In Ahlfors' argument the inner variation produces the equation in weak form

$$\text{Re} \int \int_{D'} \left(\frac{|p|^2 + |q|^2}{|p|^2 - |q|^2} \right)^m \frac{p\bar{q}}{|p|^2 + |q|^2} \partial_{\bar{z}} h \, d\xi d\eta.$$

Changing variables $z' = \zeta(z)$ we obtain

$$\text{Re} \int \int_{D \cap \{|p| > |q|\}} \left(\frac{|p|^2 + |q|^2}{|p|^2 - |q|^2} \right)^{m-1} p\bar{q} \partial_{\bar{z}} h dz \wedge d\bar{z}.$$

Set

$$U_m = \begin{cases} \left(\frac{|p|^2 + |q|^2}{|p|^2 - |q|^2} \right)^{m-1} & \text{if } |p| > |q| \\ 0 & \text{otherwise.} \end{cases}$$

and $\rho = \sum_{g' \in G'} |\partial_z g'|$ one obtains a reformulation of this PDE in terms of integration over the original surface

$$\int \int_S U_m \rho p \bar{q} \, \partial_{\bar{z}} h = 0.$$

$$\int \int_S U_m \rho p \bar{q} \, \partial_{\bar{z}} h = 0.$$

in particular this yields

Lemma 1. *The function* $f_m = U_m \rho p \bar{q}$ *is holomorphic and so describes a holomorphic quadratic differential*

$$f_m(z) dz^2$$

in D.

We let $C_m > 0$ be constants defined so that

$$U_m \rho p \bar{q} = c_m f_m(z)$$

with $\int_S |f_m| dx dy = 1$

5.6 Normal Family of Mappings with Integrable Distortion

Theorem 15. *If* $f \in W^{1,n}(\Omega, \mathbb{R}^N)$ *is a mapping whose distortion is m-integrable with* $m > n - 1$*, then* f *is continuous and the modulus of continuity depends only on the* L^m *norm of the distortion.*

In this form and in this setting, this result is due to Ahlfors [1]. It is also a consequence of work of Iwaniec and Sverak [40] and of Manfredi and Villamor [68]. See also the work of Koskela et al. [41, 42].

Given any diffeomorphism $\alpha : D \to D'$, then in view of Ascoli-Arzela and Theorem 15 one has that for every m there exists (a possibly not unique) $\zeta_m : D \to D'$ in the same homotopy class as α and which minimizes I_m. We denote by p_m, q_m its complex differentials and set

$$\min_\zeta I_m(\zeta) = I_m(\zeta_m) = \int \int_{D'} \left(\frac{|p_m|^2 + |q_m|^2}{|p_m|^2 - |q_m|^2} \right)^m.$$

Denote the quantity above by πK_m^m.

In view of Hölder inequality K_m is monotone increasing and bounded (by the dilation of α) hence $K_m \to K < \infty$. Let $0 \le \kappa < 1$ be defined by

$$K = \frac{1 + \kappa^2}{1 - \kappa^2}.$$

Normality and a diagonalization argument yields:

Lemma 2. *For a subsequence one has* $\zeta_m \to \zeta$, *as* $m \to \infty$, *uniformly on compact sets, with* ζ *quasiconformal.*

Lemma 3. *For a subsequence one has*

$$(C_m)^{\frac{1}{m}} \to K$$

as $m \to \infty$.

Lemma 4. *For a subsequence one has*

$$\int \int_D \left| |q_m| - \kappa |p_m| \right| dz \wedge d\bar{z} \to 0,$$

as $m \to \infty$.

Note that the relation

$$C_m f_m = \left(\frac{|p|^2 + |q|^2}{|p|^2 - |q|^2} \right)^{m-1} \rho(\zeta_m) p_m \bar{q}_m$$

yields

$$\frac{f_m}{|f_m|} = \frac{p_m \bar{q}_m}{|p_m||q_m|}$$

and consequently

$$\left| \frac{f_m}{|f_m|} q_m - \kappa p_m \right| = \left| \frac{p_m}{|p_m|} |q_m| - \kappa p_m \right| = \left| |q_m| - \kappa |p_m| \right|$$

Passing to a subsequence then we can assume that f_m tend to a limit f and that $\zeta_m \to \zeta$ uniformly on compact sets. The limit mapping ζ has complex derivatives p, q which are limit of p_m, q_m and thus satisfy

$$\frac{f}{|f|} q = \kappa p$$

5.7 Teichmüller Mappings in Local Parameters

A homeomorphism

$$z \to \zeta(z) : D \to D$$

whose complex derivatives satisfy

$$\kappa \frac{p}{q} = \frac{f}{|f|}$$

is called a *Teichmüller mapping*.

We show that there exists local parameters (i.e. a local set of conformal coordinates) ζ^*, z^* such that in this coordinates the map reads as the composition of two conformal transformations conjugating an affine mapping (just as in Grötzsch's problem). Denote by $\zeta \to z$ the inverse mapping and by p', q' its complex derivatives.

Differentiating the formula $z = z(\zeta(z))$ along z and $\bar{z}$ one can see that p, q, p', q' are related by the formula

$$p' = \frac{\bar{p}}{|p|^2 - |q|^2} \text{ and } q' = -\frac{q}{|p|^2 - |q|^2}.$$

If $\zeta(z)$ is quasiconformal extremal then so is $z(\zeta)$ and its associated quadratic differential $\phi(\zeta)$ satisfies:

$$\frac{\phi}{|\phi|} = \kappa \frac{p'}{q'}.$$

Consequently it follows that if $z \to \zeta$ is Teichmüller then

$$\frac{\bar{q}}{p} = \kappa \frac{\phi}{|\phi|}$$

Next we introduce two new local charts

$$z^* := \int \sqrt{f} \, dz \text{ and } \zeta^* := \int \sqrt{\zeta} \, d\zeta$$

This can be done in a sufficiently small neighborhood of a point where f, ϕ do not vanish and with fixed branches of the square roots and arbitrary integration constants.

Keeping in mind the expression $\zeta^*(\zeta(z(z^*)))$ then in terms of these new variables one has

$$p^* = \frac{d\zeta^*}{dz^*} = \frac{d\zeta^*}{d\zeta}\frac{d\zeta}{dz}\frac{dz}{dz^*} = \frac{\sqrt{\phi}}{\sqrt{f}}p$$

text and similarly

$$q^* = \frac{\sqrt{\phi}}{\sqrt{\bar{f}}}q$$

Next, observe that

$$q^* = \frac{\sqrt{\phi}}{\sqrt{\bar{f}}}q = \kappa p \frac{\bar{f}}{|f|}\frac{\sqrt{\phi}}{\sqrt{\bar{f}}} = \kappa p^* \frac{\bar{f}}{|f|}\frac{\sqrt{f}}{\sqrt{\bar{f}}} = \kappa p^*.$$

Similarly, if we use

$$\frac{\bar{q}}{p} = \kappa\frac{\phi}{|\phi|}$$

then we obtain

$$\bar{q}^* = \kappa p^*.$$

Since $q^* = \bar{q}^*$ then q^* *is real and so is* p^*.
Next, observe that

$$\frac{d}{d\bar{z}^*}(\zeta^* - \kappa\bar{\zeta}^*) = q^* - \kappa p^* = 0$$

Hence $\zeta^* - \kappa\bar{\zeta}^*$ is holomorphic and its complex derivative is

$$\frac{d}{dz^*}(\zeta^* - \kappa\bar{\zeta}^*) = p^* - \kappa q^* = p^* - \kappa\kappa p^* = p^*(1 - \kappa^2).$$

Since derivatives of a holomorphic functions are also holomorphic then p^* is both real and holomorphic, hence it must be constant.
Consequently one has that

$$\zeta^*(z^*) = p^*z^* + q^*\bar{z}^* = A(z^* + \kappa\bar{z}^*) + B$$

for some constants $A, B \in \mathbb{C}$, proving our statement on the local structure of Teichmüller mappings.

5.8   *Uniqueness (Rough Idea)*

The uniqueness part follows from a Grötzsch-like argument (more complicated in view of possible singularities). The analogues of the *rectangular regions* arise in the following way: We consider a Riemannian metric

$$ds^2 = \phi^2 d\zeta \bar{d}\zeta$$

where ϕ is the quadratic differential in the target region associated to a Teichmüller mapping. This metric is complete and non-positively curved, thus yielding unique geodesics between any pair of points. By the geodesic equation the quantity $\sqrt{\phi}d\zeta$ is constant along geodesics.

We call *horizontal arcs* those geodesics along which the argument of $\sqrt{\phi}d\zeta$ is zero. Likewise we call *vertical* those for which the argument is π.

The local charts z^*, ζ^* we have introduced earlier transforms rectangular boxes in D defined by horizontal and vertical arcs into actual rectangles in the complex plane, while at the same time the Teichmüller mapping is affine, transforming one rectangle into the other, when read in those coordinates.

An argument similar to the one we have used for Grötzsch problem yields the uniqueness and the extremality of the Teichmüller mapping.

6   A Variation on the Theme: Extremal Mappings of Finite Distortion

The integral version of the extremal mapping problem, as well as the notion of map with integrable power of the distortion have appeared in the work of Ahlfors in 1954 [1] and later in several papers from the Russian school, in particular Semenov [60–62] and references therein. As we have seen, in Ahlfors's approach to the extremal mapping problems he used a relaxation of the L^∞ variational problem, where the interest is shifted to minimizers of the L^p norm of the dilation, rather than to the L^∞ norm.

Problem 5. L^p variational problem Let $u_0 : \Omega \to \Omega'$ be a homeomorphism of finite distortion. Among all homeomorphisms $u : \Omega \to \Omega'$ whose extension to $\partial\Omega$ coincide with u_0 find one minimizing

$$\int_\Omega \psi\left(\frac{|du|}{(det(du))^{1/n}}\right)dx,$$

where $\psi : [1, \infty) \to [1, \infty)$ strictly increasing convex function with $\psi(1) = 1$.

This problem, along with generalizations to more general boundary data, has recently been studied in a sequence of papers by Astala, Iwaniecz, Martin, Onninen and several collaborators, see [7,9]. In the following we give a quick survey of their work.

Remark 9. From a Calculus of Variations point of view, one can see that following [39, Sect. 8.8.2] the functional

$$\mathscr{F}(du, \Omega) = \int_{\Omega} \psi\left(\frac{|du|}{(det(du))^{1/n}}\right) dx,$$

although not convex, is indeed *quasiconvex*, i.e. for every constant differential $A \in \mathbb{R}^{n \times n}$ and for any $\phi \in C_0^{\infty}(\Omega)$ one has

$$F(A, \Omega) \le F(A + d\phi, \Omega).$$

This notion was introduced by Morrey in 1952, see [50, 51]. Quasiconvex energy densities are those for which affine deformations are minimizers with respect to their own boundary conditions. We recall that quasiconvexity plus some growth estimates are roughly equivalent to lower semicontinuity, see Giaquinta's book [31] for a more detailed statement. Hence quasi-convexity is used often to prove existence of minimizers (as well as regularity of the extremals).

In the case at hand there are two problems:

- The growth conditions are not satisfied.
- There is a topological constraint: The space of competitors is not a vector space.

In conclusion, the results currently available from Calculus of Variations are not sufficient to attack the problem and new techniques are needed.

6.1 The Finite Distortion Version of Grötzsch Problem

Let $R = [0, 1] \times [0, 1]$ and $R' = [0, 2] \times [0, 1]$. The same argument holds for any pair of rectangles. Consider the set $\mathscr{F}=\{$ all homeomorphisms $u : R \to R'$ such that $u \in W_{loc}^{1,1}(R, \mathbb{R}^2)$ taking vertices into vertices$\}$.

Theorem 16. *[7] There is a unique minimizer for the L^1 variational problem:*

$$\min_{u \in \mathscr{F}} \int_R \left| \frac{|du|^2}{det(du)} \right|$$

Remark 10. The affine map $(x, y) \to (2x, y)$ sends R to R' by mapping vertices to vertices and has distortion

$$\frac{|du|^2}{det\, du} = \frac{5}{2}.$$

If we were to measure distortion using the operator norm we would have obtained

$$|du|_O = 2$$

and hence

$$\frac{|du|_O^2}{\det du} = 2.$$

Proof. The proof is very similar to the proof of Grötzsch problem we presented earlier: Setting $u = \alpha + i\beta$ and arguing as we did then yields

$$\int_{\partial R} \alpha \, dy = 2 \text{ and } \int_{\partial R} \beta \, dx = 1$$

Using Stokes Theorem yields

$$\int_R \alpha_x \, dxdy = 2 \text{ and } \int_R \beta_y \, dxdy = 1$$

and

$$\int_R (2\alpha_x + \beta_y) \, dxdy = 5.$$

$$5 = \int_R (2\alpha_x + \beta_y) \, dxdy \leq \int_R \sqrt{2^2 + 1} \int_R \sqrt{\alpha_x^2 + \beta_y^2} \, dxdy$$

$$\leq \sqrt{5} \int_R \|du\| \, dxdy = \sqrt{5} \int_R K(u, z) \sqrt{\det(du)} \, dxdy$$

$$\leq \sqrt{5} \sqrt{\int_R K^2(u, z) \, dxdy} \sqrt{\int_R \det(du) \, dxdy} = \sqrt{10} \sqrt{\int_R K^2(u, z) \, dxdy}.$$

This shows that the minimum of the functional is $5/2$, which is achieved by the linear map $(x, y) \rightarrow (2x, y)$. An examination of the case when the inequalities above are equalities yields that the minimum can only be achieved by this linear map. □

6.2 Trace Norm vs. Operator Norm

In [7], Astala et al. show that if the above problem one substitutes the operator norm $|A|_O = \max_{|v|=1} |Av|$ to the Hilbert-Schmidt norm, i.e. one studies minimizers of

$$\int_R \left| \frac{|du|_O^2}{\det(du)} \right|$$

then the situation changes completely and one can find infinitely many minimizers. To see this first one uses the argument above to show that

$$2 \leq \int_R \left| \frac{|du|_O^2}{\det(du)} \right|.$$

Next we observe that there is a 1-parameter family of minimizers for $a \in [0,1)$,

$$U(x,y) = \begin{cases} x + \mathbf{i}y & \text{for } x + \mathbf{i}y \in [0,a] \times [0,1] \\ \frac{2-a}{1-a}x - \frac{a}{1-a} + \mathbf{i}y & \text{for } x + \mathbf{i}y \in [a,1] \times [0,1]. \end{cases}$$

6.3 Affine Boundary Data

Following Astala, Iwaniecz, Martin and Onninen [7] we consider more general affine boundary data in higher dimension. Let us start with the case of affine orientation preserving data $u_0 : \mathbb{R}^n \to \mathbb{R}^n$ prescribed on a domain Ω with $(n-1)$ rectifiable boundary.

Theorem 17. *Given any homeomorphism of finite distortion $u : \bar{\Omega} \to \bar{\Omega}'$ such that $u = u_0$ on $\partial\Omega$ then*

$$\int_\Omega \psi\left(\frac{|du_0|^n}{\det du_0} \right) dx \leq \int_\Omega \psi\left(\frac{|du|^n}{\det du} \right) dx$$

with equality if and only if $u = u_0$ in Ω.

Sketch of the proof. We first recall two basic estimates

1. The sub-gradient inequality

$$\psi(t) - \psi(t_0) \geq \psi'(t_0)(t - t_0)$$

   valid for a.e. $t, t_0 \in [1, \infty)$.
2. The function

$$(x,y) \to x^\alpha / y^\beta$$

   defined for $x, y \in \mathbb{R}$ and $\alpha \geq \beta + 1 \geq 1$ is convex. In particular

$$\frac{x^\alpha}{y^\beta} - \frac{a^\alpha}{b^\beta} \geq \alpha \frac{a^{\alpha-1}}{b^\beta}(x - a) - \beta \frac{a^\alpha}{b^{\beta+1}}(y - \beta)$$

Using these estimates one can easily prove that

$$\psi\left(\frac{|du|^n}{det\,du}\right) - \psi\left(\frac{|du_0|^n}{det\,du_0}\right) \geq \psi'\left(\frac{|du_0|^n}{det\,du_0}\right) \frac{|du_0|^{n-2}}{det\,du_0}\langle du_0, du - du_0\rangle$$

$$+ \psi'\left(\frac{|du_0|^n}{det\,du_0}\right) \frac{|du_0|^n}{(det\,du_0)^2}(det\,du_0 - det\,du)$$

Integrating the latter over Ω yields

$$\int_\Omega \left[\psi\left(\frac{|du|^n}{det\,du}\right) - \psi\left(\frac{|du_0|^n}{det\,du_0}\right)\right] dx$$

$$\geq \int_\Omega \left[\psi'\left(\frac{|du_0|^n}{det\,du_0}\right) \frac{|du_0|^{n-2}}{det\,du_0}\langle d\,u_0, du - du_0\rangle\right.$$

$$\left. + \psi'\left(\frac{|du_0|^n}{det\,du_0}\right) \frac{|du_0|^n}{(det\,du_0)^2}(det\,du_0 - det\,du)\right] dx$$

Observe that since $du_0 = const$ and $u = u_0$ on $\partial\Omega$ then the first term on the LHS vanishes. As for the second term, note that $\int_\Omega det\,du = \int_\Omega det\,du_0 = |\Omega'|$. Thus the LHS has non-negative integral proving the first assertion. Uniqueness follow from a careful analysis of the consequences of having an identity in the above argument.

□

6.4 More General Boundary Data

The case of more general boundary data is still open. In [7], Astala et al. prove the following remarkable theorem:

Theorem 18. Let $\Omega \subset \mathbb{R}^2$ be a convex domain and set

$$\mathscr{C} = \{u \in W^{1,2}(\Omega, \mathbb{R}^2)$$

homeomorphism of finite distortion for which

$$\int_\Omega \frac{|du|^2}{det\,du} \text{ is finite}\}. \quad (12)$$

Let $u_0 \in \mathscr{C}$.

There exists a unique smooth diffeomorphism solution to the minimization problem

$$\min_{u\in\mathscr{C},u=u_0 \text{ in } \partial\Omega} \int_\Omega \frac{|du|^2}{\det du}dx$$

The key idea in the proof is to put in relation the extremal problem above with the classical Dirichlet problem

Problem 6 (n-harmonic mappings). Given $h_0 \in W^{1,n}(\Omega', \mathbb{R}^n)$, minimize the n-energy

$$\int_{\Omega'} |dh|_O^n dy$$

over the class $h \in h_0 + W_0^{1,n}(\Omega', \mathbb{R}^n)$.

The link between the two problems rests on the following theorem in [7].

Theorem 19. *Let $u \in W_{lot}^{1,n}(\Omega, \Omega')$ be a homeomorphism of finite distortion with*

$$\int_\Omega \frac{|du^{-1}|_O^n}{\det du^{-1}}(x)dx < \infty$$

The inverse map $h : \Omega' \to \Omega$ belongs to $W^{1,n}(\Omega', \Omega)$ and moreover satisfies

$$\int_{\Omega'} |dh(y)|_O^n dy = \int_\Omega \frac{|du^{-1}|_O^n}{\det du^{-1}}(x)dx.$$

See also recent developments by Hencl and Koskela [35], Hencl et al. [36, 37] and by Fusco et al. [25] and by Csörnyei et al. [19].

The proof of Theorem 19 is based on the a.e. differentiability result of Vaisala [66] and on a change of variable formula for homeomorphisms in Sobolev spaces due to Reshetnyak [58].

The two previous theorems state that the minimization problem for the n-energy

$$\int_{\Omega'} |dh|_O^2 dy$$

of $h : \Omega' \to \Omega$ in $h_0 + W_0^{1,2}(\Omega', \Omega)$ is equivalent to a minimization problem (with corresponding boundary data) for the inner distortion

$$\int_\Omega \frac{|du^{-1}|_O^2}{\det du^{-1}}(x)dx.$$

However, when $n = 2$ one has that inner and outer distortion agree, so that

$$\frac{|du|_O^2}{\det du} = \frac{|du^{-1}|_O^2}{\det du^{-1}},$$

hence minimizing the Dirichlet energy

$$\int_{\Omega'} |dh|_O^2 dy$$

is equivalent to minimizing the mean dilation

$$\int_\Omega \frac{|du|_O^2}{\det du} dx.$$

To return to the Hilbert-Schmidt norm from the operator norm we observe that in $n = 2$

$$\frac{|A|^2}{\det A} = \left(\frac{|A|_O^2}{\det A} + \frac{\det A}{|A|_O^2} \right)$$

and that the mapping

$$K \to K + \frac{1}{K}$$

is monotone. Consequently

$$\left\| \frac{|du|^2}{\det du} \right\|_\infty = \left(\left\| \frac{|du|_O^2}{\det du} \right\|_\infty + \frac{1}{\left\| \frac{|du|_O^2}{\det du} \right\|_\infty} \right)$$

then the minimization problem for the operator norm has the same solution as the one for the Hilbert-Schmidt norm. Other generalization of Grötzsch problem in higher dimension have appeared in the work of Fehlmann [23].

7 Minimal Lipschitz Extensions

We start by looking at a (relatively) simpler functional, which has been extensively studied in the last few decades.

Problem 7. Consider two sufficiently smooth bounded open sets $\Omega \subset \mathbb{R}^n$ and $\Omega' \subset \mathbb{R}^N$. Among all Lipschitz mappings $u \in Lip(\Omega, \Omega')$ with prescribed trace, find one which minimizes the functional

$$u \to \left\| du \right\|_\infty.$$

The problem is related to that of finding a canonical (unique) Lipschitz extension of the boundary map. We are interested in questions of existence, uniqueness and continuous dependence from the data.

7.1 Aronsson's Approach in the Scalar Case $N = 1$

In the following we describe the $N = 1$ scalar case for this L^∞ variational problem. Since $||\nabla u||_\infty$ is equivalent to the Lipschitz norm of the scalar function $u : \Omega \to \mathbb{R}$ this leads to the following

Definition 12. *A minimizing Lipschitz extension* is an extension of a Lipschitz scalar function $f : \partial\Omega \to \mathbb{R}$ to $u : \Omega \to \mathbb{R}$ with $u = f$ on $\partial\Omega$ and

$$Lip(u, \Omega) = \sup_{x \neq y, x, y \in \Omega} \frac{|u(x) - u(y)|}{|x - y|} = Lip(f, \partial\Omega)$$

In 1934, independently E. J. MacShane [49] and H. Whitney [69] noted the following:

Theorem 20. *Such extensions always exist but are not unique.*

The proof of existence is based on the following observation: Assume that an extension u exists and let $\lambda = Lip(f, \partial\Omega)$. Since $Lip(u, \Omega) = Lip(f, \partial\Omega)$ then for all $x \in \partial\Omega$ and all $y \in \Omega$ one must have

$$-\lambda \leq \frac{|u(y) - f(x)|}{d(x, y)} \leq \lambda,$$

and hence

$$f(x) - \lambda d(x, y) \leq u(y) \leq f(x) + \lambda d(x, y).$$

Since for all $x \in \partial\Omega$ and $y \in \Omega$

$$f(x) - \lambda d(x, y) \leq u(y) \leq f(x) + \lambda d(x, y).$$

if we define the upper and lower functions

$$L(y) = \sup_{x \in \partial\Omega} (f(x) - \lambda d(x, y)) \text{ and } U(y) = \inf_{x \in \partial\Omega} (f(x) + \lambda d(x, y))$$

then these are minimizing Lipschitz extensions of f and so is any u such that $L \leq u \leq U$ in Ω.

Remark 11. We recall an example due to Jensen [44], showing that the problem of minimal Lipschitz extension does not have a unique solution. Let $\Omega = B(0, 1)$ and $f(x, y) = 2xy$. For every $0 \leq \alpha \leq 1/2$ set

$$u^\alpha(x, y) = \begin{cases} 0 & \text{for } x^2 + y^2 \leq \alpha^2 \\ \frac{2xy(\sqrt{x^2+y^2}-\alpha)}{(1-\alpha)(x^2+y^2)} & \text{for } \alpha^2 \leq x^2 + y^2 \leq 1. \end{cases}$$

Note that for $x^2 + y^2 = 1$ we have $u^\alpha(x, y) = 2xy$ and more over

$$Lip(u^\alpha, \Omega) = Lip(f, \partial\Omega)$$

So there are infinitely many distinct minimal Lipschitz extensions.

Problem 8. Is there a special class of *canonical* extensions for which uniqueness holds?

In 1967 G. Aronsson (see [4]) proposed a way to *localize* the functional by introducing the formal approximation scheme:

• Consider minimzers u_p of

$$\int |\nabla u|^p$$

They are *p-harmonic*, i.e. weak solutions of the equation

$$\Delta_p u_p = div(|\nabla u_p|^{p-2}\nabla u_p) = 0$$

• In case $u \in C^2$ then we can rewrite this PDE in non-divergence form

$$(p - 2)|\nabla u|^{p-4}\left(u_i u_j u_{ij} + \frac{|\nabla u|^2}{p - 2}\Delta u\right) = 0$$

• Let $p \to \infty$ and formally obtain the ∞-*Laplacian*

$$\Delta_\infty u = u_{ij}u_i u_j = \frac{1}{2}\langle\nabla|\nabla u|^2, \nabla u\rangle = 0.$$

Remark 12. Note that a priori there is no link between solutions of the non-linear, degenerate elliptic PDE

$$u_i u_j u_{ij} = 0$$

and the problem of minimal Lipschitz extensions. *The previous computation is purely formal.*

In [4], Aronsson established a link between sufficiently smooth solutions of the infinity Laplacian and correspondingly smooth minimizers of the L^∞ variational problem.

Theorem 21 (∞-harmonic implies AMLE [4]). *C^2 solutions of $\Delta_\infty u = 0$ are Absolute Minimizing Lipschitz Extensions (AMLE), i.e. they minimize $Lip(u, D)$ on every subdomain $D \subset \Omega$*

$$Lip(u, D) = Lip(u, \partial D) \text{ for every } D \subset \Omega$$

In some sense, the localization built-in in the notion of AMLE is inherited from the L^p problem. The key observation in the proof is that for C^2 solutions one has $|\nabla u|$ is constant along integral curves of ∇u in $D \subset \Omega$. Aronsson proved that such curves cannot vanish in the interior of the domain and cannot wind up infinitely many times within the domain, hence they have to reach the boundary.

As a converse to the previous theorem, Aronsson also proved

Theorem 22. *Every C^2 AMLE is ∞-harmonic.*

Regarding existence of AMLE, Aronsson established the following

Theorem 23 (Existence of AMLE). *Given any $\Omega \subset \mathbb{R}^n$ and $f \in Lip(\partial\Omega)$ there exists always a AMLE.*

We say that a minimal Lipschitz extension $u \in Lip(\Omega)$, with boundary values $f \in Lip(\partial\Omega)$, has the property $\mathscr{A}$, if for every $D' \subset D$ one has $u \leq U'$ in D' where U' is the upper function in D' with respect to the boundary value u.

The AMLE corresponding to $f \in Lip(\partial\Omega)$ is then defined as

$$u(x) := \inf_g g(x)$$

where the inf is taken over all functions with the $\mathscr{A}$ property with respect to f.

In 1968 Aronsson proved that there can be at most one $u \in C^2(\Omega) \cap C(\bar{\Omega})$ solution of the ∞-Laplacian. Thus showing that there can be at most one C^2 AMLE.

Remark 13. Aronsson shows that C^2 solutions have nowhere vanishing gradient, however any C^2 solution with boundary data $2xy$ must have a critical point. So there may not be C^2 solutions of the ∞-Laplacian for this data.

The C^2 hypothesis in Aronsson's work was a severe limitation until in 1993 Jensen (see [44]) removed it using the theory of viscosity solutions, and eventually proving *uniqueness of AMLE.*

Definition 13 (Viscosity solution). A continuous function u is ∞-subharmonic in the viscosity sense if for any point $x \in \Omega$ and $\phi \in C^2(\Omega)$ such that $\phi - u$ has a minimum at x one has $\phi_{ij}\phi_i\phi_j \geq 0$. Supersolutions and solutions are defined in a similar fashion.

See [5, 17, 18] for a broad exposition and a detailed list of references.

Theorem 24 (Bhattacharya et al. [15]). *Given fixed boundary data, p-harmonic functions converge to viscosity ∞-harmonic functions as $p \to \infty$.*

Theorem 25 (Jensen [44]). *AMLE are viscosity ∞-harmonic functions.*

Theorem 26 (Jensen [44]). *The Dirichlet problem for viscosity ∞-harmonic functions has a unique solution.*

At present, thanks to the work of Armstrong, Barron, Champion, Crandall, De Pascale, Evans, Gariepy, Jensen, Juutinen, Manfredi, Oberman, Parviainen, Rossi, Smart, Wang, Yu (to quote just a few) as well as the more recent approach of Naor, Peres, Sheffield, Schramm and Wilson one can prove the uniqueness of AMLE in a variety of ways. In particular, this can be achieved without directly using the ∞-Laplacian operator and viscosity solutions for PDE. See [5, 17] for a detailed account of these developments.

However, at present, out of this multitude of approaches there is no method that can be immediately extended to approach *uniqueness* in the vector valued case.

7.2 Aronsson's Approach in the Vector-Valued Case $N > 1$

Existence of minimizing Lipschitz extensions for mappings follows from the classical

Theorem 27 (Kirszbraun's Theorem). *Let X, Y be two Hilbert spaces and $U \subset X$ and open set. If $f : U \to Y$ is a Lipschitz mapping then there exits an extension $F : X \to Y$ with the same Lipschitz constant.*

However, as we have seen, such extensions may not be unique. Generalizing Aronsson's approach beyond the real-valued mappings setting, and in particular to the vector-valued case $u : \Omega \to \mathbb{R}^N$, is very challenging but, aside from being an important problem in its own right, may have several potential applications in image processing (specifically image inpainting and surface reconstruction).

A first step in this direction was taken by Naor and Sheffield in [52] where the focus is on absolutely minimizing Lipschitz extensions in the context of tree-valued mappings. Their main result in [52] consists in existence of a unique AMLE of any prescribed Lipschitz mappings from a subset of a locally compact length metric space to a metric theory. Among other things, the authors also introduce a general definition of discrete infinity harmonic function and prove existence of infinity harmonic extensions.

Shortly afterwards, in [63], Sheffield and Smart considered minimizing extensions of the Lipschitz norm

$$Lip(u, \Omega) = \sup_{x,y \in \Omega} \frac{d(u(x), u(y))}{d(x, y)} = \sup_\Omega |du|_O.$$

as well as its *discrete* analogue for mappings $u : G \to R^N$ where $G = (E, X, Y)$ is a finite graph

$$Su(x) := \sup_{y \approx x} d(u(x), u(y))$$

and two vertices $x, y \in E$ are in relation if they are separated by an edge. The subset of vertices $Y \subset X$ here plays the role of the domain for the mapping to be extended.

Definition 14. A mapping is said to be discrete ∞-harmonic at $x \in X \setminus Y$ if there is no way to decrease $Su(x)$ by changing the value of u at x.

Peres, Schramm, Sheffield and Wilson have shown that for any Lipschitz $f : Y \to \mathbb{R}$ there exists a unique Lipschitz extension $u : X \to \mathbb{R}$ which is ∞-harmonic. In [63], Sheffield and Smart prove that the uniqueness fails for the vector valued case.

To recover uniqueness in [63] Sheffield and Smart introduce a new notion, that of *tight extension* that is stronger than discrete ∞-harmonic:

Definition 15 (Tightness). Consider mappings $u, v : X \to \mathbb{R}^N$ that agree on Y. The mapping v is tighter that u on G if

$$\sup\{su|\ su > Sv\} > \sup\{Sv|\ Sv > su\}.$$

The mapping u is *tight* on G if there is no tighter v.

Theorem 28 (Sheffield and Smart [63]). *Let $G = (E, X, Y)$ be a finite connected graph.*

- *Every Lipschitz $f : Y \to \mathbb{R}^N$ has a unique tight extension $u : X \to \mathbb{R}^N$. Moreover u is tighter than every other extension of f.*
- *For every $p > 0$ consider a minimizer $u_p : X \to Y$ of $I_p(w) := \sum_x (Sw(x))^p$, with $u_p = f$ on Y. As $p \to \infty$ the mappings $u_p \to u$ pointwise, where u is the tight extension of f.*

Motivated by this result, Sheffield and Smart introduced a notion of tight extension in the continuous setting: First one sets $Lu(x) = \inf_{r>0} Lip(u, \Omega \cap B(x, r))$.

Definition 16. Let $u, v \in C(\bar{\Omega}, \mathbb{R}^N)$ be two Lipschitz function which agree on $\partial\Omega$. We say that v is tighter that u if

$$\sup\{Lu|\ Lv < Lu\} > \{Lv|\ Lv > Lu\}.$$

A mapping u is called *tight* if there is no tighter v.

Definition 17. A principal direction for a mapping $u \in C^1(\Omega, \mathbb{R}^N)$ is a continuous, unit vector field in Ω such that at each point it spans the principal eigenspace of

$du^T du$. If $N = 1$ then the field is $-\nabla u/|\nabla u|$. Note that the existence of a principal direction field implies that the largest eigenvalue for $du^T du$ is simple.

Recall that the linear transformation $y \to du(x)y$ sends spheres into ellipsoids. The principal direction corresponds to the largest axis of such ellipsoid.

Theorem 29 (Sheffield and Smart [63]). *Let $u \in C^3(\Omega, \mathbb{R}^N)$ have a principal direction field $a \in C^2(\Omega, \mathbb{R}^n)$. The mapping u is tight if and only if*

$$(u^i_j a_j)_k a_k = 0.$$

Theorem 30 (Sheffield and Smart [63]). *Let $\Omega \subset \mathbb{C}$ be a bounded open set and $u : \Omega \to \mathbb{C}$ be analytic in a neighborhood of $\bar{\Omega}$. The mapping u is tight if and only if either*

(i) $\partial_z \partial_z u = 0$ in Ω (i.e., u is affine); or
(ii) The meromorphic function

$$Re\left(\frac{u_z u_{zz}}{(u_{zzz})^2}\right) \le 2,$$

in the set where $u_{zz} \ne 0$.

If u is a diffeomorphism and u_{zz} never vanishes then part (ii) can be rewritten as $(\Delta - \Delta_\infty) \log |u_z^{-1}| \le 0$. In other words, the level sets of $|u_z^{-1}|$ are convex.

7.3 A Refinement of the Aronsson Equation

If we use Aronsson's scheme in the scalar case then we have seen as (with sufficient regularity) the approximating p-harmonic functions satisfy

$$(p - 2)|\nabla u|^{p-4}\left(u_i u_j u_{ij} + \frac{|\nabla u|^2}{p - 2}\Delta u\right) = 0$$

In the vector case, using the Euclidean norm this time, it is easy to see that one obtains instead

$$du_{lk}\partial_{jk}u^l \partial_j u + \frac{|du|^2\Delta u}{p - 1} = 0. \tag{13}$$

If one lets $p \to \infty$ then formally we obtain the ∞-Laplacian system

$$u^i_{jk}u^l_k u^i_j = 0 \text{ for } i = 1,\ldots,N.$$

$$u_{jk}^l u_k^l u_j^i = 0 \text{ for } i = 1, \dots, N.$$

Theorem 31 (Katzourakis [47]). *There exists distinct solutions of the ∞-Laplacian above with the same boundary data.*

The explicit counterexamples are all 1-dimensional, with $\Omega \subset \mathbb{R}$. In view of such examples it appears that the ∞-Laplacian analogue may not an appropriate PDE to characterize unique extremals. In [47], Katzourakis observed that one can recover more information, leading to an augmented (formal) Aronsson system: Recall

$$du_{lk}\partial_{jk}u^l\partial_j u + \frac{|du|^2 \Delta u}{p-1} = 0.$$

Notice that the term $du_{lk}\partial_{jk}u^l\partial_j u$ lies in the image of du. Consequently for (13) to hold we must also have

$$\pi_{N(du)}\Delta u = 0$$

where $N(du) = \{v \in \mathbb{R}^N \mid duv = 0\}$ is the null-space of the linear application $v \to duv$ and π_N denotes the orthogonal projection in $\mathbb{R}^N$ onto such space. Thus a more complete choice for the Aronsson system would be the coupled system

$$u_{jk}^l u_k^l u_j^i = 0 \text{ and } \pi_{N(du)}\Delta u = 0, \tag{14}$$

As noted in [47], this system may have discontinuous coefficients even for smooth du, since the rank of du may change from point to point. Although the previous derivation is purely formal one has the following variational interpretation

Theorem 32 (Katzourakis [47]). *Let $\Omega \subset \mathbb{R}^n$ and $u \in C^2(\Omega, \mathbb{R}^n)$ be diffeomorphism with non-vanishing Jacobian. The mapping u solves (14) if and only for every subdomain $D \subset\subset \Omega$ and for every $g \in Lip_0(D, \mathbb{R})$ and $\xi \in \mathbb{R}^n$ one has*

$$||\nabla u||_{L^\infty(D)} \le ||\nabla(u + g\xi)||_{L^\infty(D)}$$

The actual result is more general and involves C^2 mappings $u : \Omega \to \mathbb{R}^N$ and an additional variational characterization.

8 Aronsson's Approach for the Extremal Dilation Problem

In this final section we return to the extremal problem for quasiconformal mappings and recall recent results by Raich and the author [16] in which the Aronsson's approximation scheme is used to introduce a notion of absolute minimizers in the quasiconformal setting. The goal here is to find a candidate PDE that would play the role similar to that of the infinity-Laplacian in the AMLE theory. Following the approach of Sheffield and Smart in [63] we focus on the C^2 case. Although this is an unnatural smoothness hypothesis for quasiconformal mappings, it does provide some insights into the general problem.

The first step in this approach consists in studying extremal mappings for the corresponding L^p problem. If $p > 1$, $\Omega \subset \mathbb{R}^n$ and the diffeomorphism $u \in C^2(\Omega, \mathbb{R}^n)$ is a critical point of the functional

$$\mathscr{F}_p(u, \Omega) := \int \frac{|du|^{np}}{(detdu)^p} dx \tag{15}$$

then the mapping u satisfies the system of Euler-Lagrange equations

$$(L_p u)^i = np\partial_j \left(\left[trace(g) \right]^{\frac{np-2}{2}} du^{-1,T} S(g) \right)_{ij}, \text{ for } i = 1, \ldots, n$$

If we let $p \to \infty$ then formally one obtains the Aronsson PDE,

$$(L_\infty u)^i = S(g)_{ij}\partial_j \sqrt{trace(g)} = 0 \text{ for } i = 1, \ldots, n. \tag{16}$$

This PDE tells us that the *dilation of the mapping u (i.e. $\sqrt{trace(g)}$) is constant along curves tangent to the sub-bundle generated by the rows of $S(g)$.*

Problem 9. What is the lowest regularity for the mapping u for which the PDE

$$(L_\infty u)^i = S(g)_{ij}\partial_j \sqrt{trace(g)} = 0 \text{ for } i = 1, \ldots, n$$

is meaningful?

Remark 14. It is tempting to define solutions of (16) as quasiconformal mappings such that their dilation $trace(g)$ is constant along all curves tangent to the sub-bundle generated by the rows of $S(g)$. Observe that for this definition to be meaningful at the very least one would need regularity for u such that constant linear combinations of the rows of $S(g)$ generate integral flows (for instance $S(g) \in BV$) and the quantity $trace(g)$ must be continuous (so it can be evaluated along such integral curves).

It is important to note that classical solutions of the extremal quasiconformal problems, e.g. Teichmüller mappings, solve (16) in the regions where they are C^2 smooth.

Proposition 4. *(1) Any Teichmüller map of the form $u := \psi \circ v \circ \phi^{-1}$ with ψ, ϕ conformal and v affine is a solution of $L_\infty u = 0$. (2) the quasiconformal mappings $u(x) = |x|^{\alpha-1}x$ for $\alpha > 0$ solve $L_\infty u = 0$ away from the origin. (3) Let $0 < \alpha < 2\pi$ and (r, θ, z) be cylindrical coordinates for $x = (x_1, \ldots, x_n)$ where $x_1 = r\cos\theta$, $x_2 = r\sin\theta$ and $x_j = z_j$, $3 \le j \le n$. The quasiconformal mapping*

$$u(r, \theta, z) = \begin{cases} (r, \pi\theta/\alpha, z) & 0 \le \theta \le \alpha \\ (r, \pi + \pi\frac{\theta-\alpha}{2\pi-\alpha}, z) & \alpha < \theta < 2\pi \end{cases} \tag{17}$$

solves $L_\infty u = 0$ away from the set $r = 0$.

The following theorem establishes a link between the formal PDE (16) and the L^∞-variational problem.

Theorem 33 ([16]). *If $u : \Omega \to \mathbb{R}^n$ C^2 is a quasiconformal solution of $L_\infty u = 0$ in Ω, then*

(i) *For every $D \subset \Omega$ one has $\sup_{\bar{D}} \sqrt{trace(g)} = \sup_{\partial D} \sqrt{trace(g)}$.*
(ii) *For every $D \subset \Omega$ one has $\inf_{\bar{D}} \sqrt{trace(g)} = \inf_{\partial D} \sqrt{trace(g)}$.*
(iii) *There exists $C = C(n) > 0$ such that for every C^2 domain $D \subset \Omega$ and $w : \bar{D} \to \mathbb{R}^n$ C^2 quasiconformal such that $u = w$ on ∂D one has $\sup_D \sqrt{trace(g(u))} \le C \sup_D \sqrt{trace(g(w))}|$.*
(iv) *If $n = 2$ the dilation $|g|$ is constant in Ω and if u is affine in a neighborhood of $\partial\Omega$ then u must be an affine transformation throughout Ω.*

Sketch of the proof. Show that any interior maximum points for $|g|$ propagate along curves tangent to the span of the rows of $S(g)$ until they reach the boundary. This is achieved by using the fact that $S(g)$ is either vanishing or has at least rank higher than two. This is used to construct a non self-intersecting curve of this kind and showing that (i) its total length must be finite; (ii) the curve cannot vanish in Ω. Points (i) and (ii) imply then that the curve must reach the boundary. $\qquad\square$

Theorem 34 ([16]). *If $u : \Omega \to \mathbb{R}^n$ C^2 is a quasiconformal absolute extremal, i.e. for every $D \subset \Omega$ and $w : \bar{D} \to \mathbb{R}^n$ C^2 quasiconformal such that $u = w$ on ∂D one has $\sup_D \sqrt{trace(g(w))} \le \sup_D \sqrt{trace(g(w))}$, then $L_\infty u = 0$ in Ω.*

Sketch of the proof. Arguing by contradiction we assume there is a ball $B \subset\subset \Omega$ s.t. $L_\infty u \ne 0$ in B. We construct a *better competitor* for the variational problem: i.e. a C^2 quasiconformal diffeomorphism $V : \bar{B} \to \mathbb{R}^n$ with same boundary values as u on ∂B and $\sup_B trace\, g(V) < \sup_B trace\, g(u)$. This is done by perturbing u with a finite number of "bumps" that reduce the dilation near the boundary. $\qquad\square$

Remark 15. A similar result was proved much earlier by Barron et al. in their important work [12] with a different, less constructive proof. The advantage of the approach in [16] is that it provides a competitor which is also quasiconformal.

Remark 16. Recently in [46], Katzourakis applied the refined derivation technique we described earlier to the quasiconformal setting and obtained the formal extended system:

$$du_{ak}J_{ki}du_{bl}J_{lj}\partial_{kl}u^b + |du|^2[\pi_{N(duJ)}]_{ab}J_{ij}\partial_{ij}u^b = 0$$

where $J = g^{-1}S(g)$ and $g = du^T du$. The equation is composed of two linearly independent parts. The first, in the case of diffeomorphisms between domains of $\mathbb{R}^n$

coincides with the system we have described earlier. The second component is new but it is not yet clear how it relates to the variational problem. The paper [46] also provides a necessary and sufficient condition for C^2 mappings to satisfy this system.

8.1 A Gradient Flow Approach

Let $\Omega \subset \mathbb{R}^n$ is a bounded, $C^{2,\alpha}$ smooth, open set. Currently we do not know how to prove existence of solutions of (16) or how to attack the extremal problems for a fixed homotopy class of quasiconformal mappings. A possible strategy for a proof consists in finding solutions of a gradient flow $u_p(x, t)$ for the functional $\mathscr{F}_p(u, \Omega)$ defined in (15). The long term existence and suitable estimates (independent of p as $p \to \infty$) for such flow then would yield the existence of the asymptotic mapping $w_p(x) = \lim_{t \to \infty} u_p(x, t)$ which would be a candidate for the L^p minimization problem within the homotopy class of the initial data. The solution to the L^∞ problem then could be achieved by establishing estimates on w_p independent of p and letting $p \to \infty$.

For a fixed diffeomorphism $u_0 : \Omega \to \mathbb{R}^n$, we want to study diffeomorphism solutions $u(x, t)$ of the initial value problem

$$\begin{cases} \partial_t u = -L_p u & \text{in } \Omega \times (0, T). \\ u = u_0 & \text{at } \Omega \times \{t = 0\} \end{cases} \tag{18}$$

where we recall that

$$(L_p u)^i = np\partial_j \left(|g|^{\frac{np-2}{2}} du^{-1,T} S(g) \right)_{ij}, \quad \text{for } i, j = 1, \ldots, n$$

If there is a $T > 0$ such that a solution $u \in C^2(\Omega \times (0, T))$ exists with $\det du > 0$ in $\Omega \times (0, T)$, then

$$\frac{d}{dt}\mathscr{F}_p(u, \Omega) = -\left(\frac{1}{|\Omega|} \int_\Omega |L_p u|^2 dx \right) \leq 0,$$

i.e., the p-distortion is nonincreasing along the flow. Hence we obtain

Proposition 5. *If $u \in C^2(\Omega \times [0, T), \mathbb{R}^n) \cap C^1(\bar{\Omega} \times [0, T), \mathbb{R}^n)$ is a solution of (18) with $\det du > 0$ in $\bar{\Omega} \times [0, T)$, then for all $0 \leq t < T$, $\|\mathbb{K}_{u_p}\|^p_{L^p(\Omega)} = \|\mathbb{K}_u\|^p_{L^p(\Omega)} - \int_0^T \|L_p u(\cdot, t)\|_{L^2(\Omega)} dt$ and consequently*

$$\|\mathbb{K}_u\|_{L^p(\Omega \times \{t\})} \leq \|\mathbb{K}_{u_0}\|_{L^p(\Omega)}. \tag{19}$$

It is immediate to show that the functional $\mathscr{F}_p(u, \Omega)$ is invariant by conformal deformation. Therefore, if we let $s \mapsto F_s : \mathbb{R}^n \to \mathbb{R}^n$ be a one-parameter semi-group of conformal transformations, then solutions to the PDE system

$$\partial_t u = L_p u + \frac{d}{ds} F_s(u)\bigg|_{s=0}$$

would also satisfy (19). Recall that the flow F_s is conformal if

$$S(d\mathscr{D}) = \frac{d\mathscr{D} + d\mathscr{D}^T}{2} - \frac{1}{n}\,\text{trace}\,(d\mathscr{D})I_n = 0$$

where $\mathscr{D} = \left(\frac{d}{ds}F_s\right)\bigg|_{s=0} \circ F_0^{-1} = \left(\frac{d}{ds}F_s\right)\bigg|_{s=0}$ and S denotes the Ahlfors operator.
If $n = 2$ then this amounts to $\partial_{\bar{z}}\mathscr{D} = 0$. If $n \geq 3$ there is more rigidity and conformality requires that

$$\mathscr{D}(x) = a + Bx + 2(c \cdot x)x - |x|^2 c$$

for any vectors a, c and matrix B with $S(B) = 0$ (see [59]).

We observe that in light of conformal invariance, if $u(x,t)$ is a solution of (18) in $\Omega \times (0,T)$ and $v(x,t) = \delta u(\lambda x, \delta^{-2} t)$ for some $\lambda, \delta > 0$, then $v(x,t)$ is still a solution with initial data $v_0(x) = \delta u_0(\lambda x)$ in an appropriately scaled domain. Applying inversions in a suitable way will also yield new solutions from $u(x,t)$.

Usual elliptic/parabolic PDE techniques do not apply. The main difficulty consists in the fact that the functional is *not convex* but only quasi-convex (in the sense of Morrey). In order to study the gradient flow it helps to rewrite the system in non-divergence form.[3]

$$(L_p u)^i = A^{ik}_{j\ell}(du)u^k_{j\ell}.$$

with

$$A^{ik}_{j\ell}(q) = -p\frac{|q|^{np-2}}{(\det q)^p}\left[np(q_{k\ell}q^{ji} + q_{ij}q^{\ell k}) - n(np-2)\frac{q_{ij}q_{k\ell}}{|q|^2} - |q|^2(q^{\ell i}q^{jk} + pq^{\ell k}q^{ji}) - n\delta_{ki}\delta_{j\ell}\right].$$

This form of the PDE has a remnant of ellipticity in the form of the so-called *Legendre-Hadamard* property: There exists constants $C_1, C_2 > 0$ depending respectively only on n and on p and on n such that for a.e. $q \in \mathbb{R}^{n \times n}$ and for all $\xi, \eta \in \mathbb{R}^n$

$$C_1(n,p)p|\eta|^2|\xi|^2\frac{|q|^{np-2}}{(\det q)^p} \leq A^{ik}_{j\ell}(q)\eta_i\xi^j\eta_k\xi^\ell$$

$$\leq C_2(n)p^2|\eta|^2|\xi|^2\left(\frac{|q|^{np-2}}{(\det q)^p} + \frac{|q|^{n(p+2)-2}}{(\det q)^{p+2}}\right)$$

---

[3]To do this however one has to assume existence of two derivatives for the solution.

Using the latter, Raich and the author established in [16] certain Schauder type estimates for the gradient flow (i.e. a gain of two derivatives with respect to the regularity of the right hand side and the coefficients of the PDE). The Schauder estimates in turn allow to rephrase the system (18) as a fixed point problem for a contraction map, leading to the short time existence and uniqueness result

Definition 18. Let $\Omega \subset \mathbb{R}^n$ be a smooth bounded domain and for $T > 0$ let $Q = \Omega \times (0, T)$. The *parabolic boundary* is defined by $\partial_{par} Q = (\Omega \times \{t = 0\}) \cup (\partial\Omega \times (0, T))$. The *parabolic distance* is $d((x, t), (y, s)) := \max(|x - y|, \sqrt{|t - s|})$. For $\alpha \in (0, 1)$ we define the *parabolic Hölder class* $C^{0,\alpha}(Q) := \{u \in C(Q, \mathbb{R}) | \; \|u\|_{C^\alpha(Q)} := [u]_\alpha + \|u\|_0 < \infty\}$, where

$$[u]_\alpha := \sup_{(x,t),(y,s)\in Q \text{ and } (x,t)\neq(y,s)} \frac{|u(x,t) - u(y,s)|}{d((x,t),(y,s))^\alpha}$$

and $|u|_0 = \sup_Q |u|$. For $K \in \mathbb{N}$ we let $C^{K,\alpha}(Q) = \{u : Q \to \mathbb{R}| \; \partial_{x_{i_1}} \cdots \partial_{x_{i_K}} u \in C^{0,\alpha}(Q)\}$.

Proposition 6. *Let $u_0 : \Omega \to \mathbb{R}^n$ be a $C^{2,\alpha}$ diffeomorphism for some $0 < \alpha < 1$ with $\det du_0 \geq \varepsilon > 0$ in $\bar\Omega$. Assume that $L_p u_0 = 0$ for all $x \in \partial\Omega$. There exist constants $C, T > 0$ depending on $p, n, \Omega, \varepsilon, \|u_0\|_{C^{1,\alpha}(\bar\Omega)}$, and a sequence of diffeomorphisms $\{u^h\}$ in $C^{2,\alpha}(Q)$ with $Q = \Omega \times (0, T)$ so that*

(a) $\det u^h \geq \frac{\varepsilon}{2}$ *for all $(x, t) \in Q$,*
(b) $\|u^h\|_{C^{2,\alpha}(Q)} + \|\partial_t u^h\|_{C^{0,\alpha}(Q)} \leq C \|u_0\|_{C^{2,\alpha}(\Omega)}$,
(c) $\begin{cases} \partial_t u^{h,i} - A_{jl}^{ik}(du^{h-1})\partial_j \partial_l u^{h,k} = 0 & \text{in } Q \\ u^h = u_0 & \text{on } \partial_{par} Q. \end{cases}$

Theorem 35. *If $u(x, 0) \in C^{2,\alpha}$+boundary conditions then there exists a unique $C^{2,\alpha}_1(\Omega \times (0, T), \mathbb{R}^n)$ solutions for small $T = T(p, n, u_0, \Omega) > 0$.*

Although the previous result establishes short time existence, the dependence of the interval of existence from p remains an obstacle to the study of the asymptotic limit $p \to \infty$. In order to carry out the program we outlined earlier, one would need a global existence result, as well as estimates independent of p as $p \to \infty$. Currently there is very little literature about gradient flows of quasi-convex functionals but a an important paper of Evans et al. [21] lays out a strategy to obtain global estimates: Following [21], Raich and the author in [16] let $\beta = \det du^{-1}$ then show that β solves the scalar PDE

$$\partial_t \beta = [a_{ij}(du)\beta]_{ij}$$

with

$$a_{ij} = p\left(\delta_{ij} - n\frac{du_{jk}du_{ik}}{|du|^2}\right)\sqrt{|g|}^{np}.$$

Although the lack of a sign in the symbol prevents us from using the maximum principle and establishing immediate global bounds, this PDE is a starting point for the study of global estimates.

Acknowledgements We wish to thank C. Gutierrez and E. Lanconelli for the scientific organization of the C. I. M.E. course and for inviting the author to present these lectures. We are also grateful to P. Zecca and to all the staff of C. I.M.E. for their logistic support and hospitality. The author is partially supported by the US National Science Foundation through grants DMS-1101478 and DMS-0800522

References

1. L.V. Ahlfors, On quasiconformal mappings. J. Analyse Math. **3**, 1–58 (1954); correction, 207–208
2. L.V. Ahlfors, An introduction to the theory of analytic functions of one complex variable, in *Complex Analysis*, 3rd edn. International Series in Pure and Applied Mathematics (McGraw-Hill, New York, 1978)
3. L.V. Ahlfors, *Lectures on Quasiconformal Mappings*, 2nd edn. University Lecture Series, vol. 38 (American Mathematical Society, Providence, 2006). With supplemental chapters by C. J. Earle, I. Kra, M. Shishikura and J. H. Hubbard
4. G. Aronsson, Extension of functions satisfying Lipschitz conditions. Ark. Mat. **6**(1967), 551–561 (1967)
5. G. Aronsson, M.G. Crandall, P. Juutinen, A tour of the theory of absolutely minimizing functions. Bull. Am. Math. Soc. (N.S.) **41**(4), 439–505 (2004)
6. O.A. Asadchii, On the maximum principle for n-dimensional quasiconformal mappings. Mat. Zametki **50**(6), 14–23, 156 (1991)
7. K. Astala, T. Iwaniec, G.J. Martin, J. Onninen, Extremal mappings of finite distortion. Proc. Lond. Math. Soc. (3) **91**(3), 655–702 (2005)
8. K. Astala, T. Iwaniec, G. Martin, *Elliptic Partial Differential Equations and Quasiconformal Mappings in the Plane*. Princeton Mathematical Series, vol. 48 (Princeton University Press, Princeton, 2009)
9. K. Astala, T. Iwaniec, G. Martin, Deformations of annuli with smallest mean distortion. Arch. Ration. Mech. Anal. **195**(3), 899–921 (2010)
10. Z.M. Balogh, K. Fässler, I.D. Platis, Modulus of curve families and extremality of spiral-stretch maps. J. Anal. Math. **113**, 265291 (2011)
11. E.N. Barron, R.R. Jensen, C.Y. Wang, Lower semicontinuity of L^∞ functionals. Ann. Inst. H. Poincaré Anal. Non Linéaire **18**(4), 495–517 (2001)
12. E.N. Barron, R.R. Jensen, C.Y. Wang, The Euler equation and absolute minimizers of L^∞ functionals. Arch. Ration. Mech. Anal. **157**(4), 255–283 (2001)
13. P. Bauman, D. Phillips, N.C. Owen, Maximal smoothness of solutions to certain Euler-Lagrange equations from nonlinear elasticity. Proc. Roy. Soc. Edin. Sect. A **119**(3–4), 241–263 (1991)
14. L. Bers, Quasiconformal mappings and Teichmüller's theorem, in *Analytic Functions* (Princeton University Press, Princeton, 1960), pp. 89–119
15. T. Bhattacharya, E. DiBenedetto, J. Manfredi, Limits as $p \to \infty$ of $\Delta_p u_p = f$ and related extremal problems. Rend. Sem. Mat. Univ. Politec. Torino, Special Issue (1989), 15–68 (1991). Some topics in nonlinear PDEs (Turin, 1989)
16. L. Capogna, A. Raich, *An Aronsson-Type Approach to Extremal Quasiconformal Mappings in Space*. Preprint (2010)

17. M.G. Crandall, A visit with the ∞-Laplace equation, in *Calculus of Variations and Nonlinear Partial Differential Equations*. Lecture Notes in Mathematics, vol. 1927 (Springer, Berlin, 2008), pp. 75–122
18. M.G. Crandall, H. Ishii, P.L. Lions, User's guide to viscosity solutions of second order partial differential equations. Bull. Am. Math. Soc. (N.S.) **27**(1), 1–67 (1992)
19. M. Csörnyei, S. Hencl, J. Malý, Homeomorphisms in the Sobolev space $W^{1,n-1}$. J. Reine Angew. Math. **644**, 221–235 (2010)
20. B. Dacorogna, W. Gangbo, Extension theorems for vector valued maps. J. Math. Pure Appl. (9) **85**(3), 313–344 (2006)
21. L.C. Evans, O. Savin, W. Gangbo, Diffeomorphisms and nonlinear heat flows. SIAM J. Math. Anal. **37**(3), 737–751 (2005) (electronic)
22. D. Faraco, X. Zhong, Geometric rigidity of conformal matrices. Ann. Sc. Norm. Super. Pisa Cl. Sci. (5) **4**(4), 557–585 (2005)
23. R. Fehlmann, Extremal problems for quasiconformal mappings in space. J. Analyse Math. **48**, 179–215 (1987)
24. G. Friesecke, S. Müller, R.D. James, Rigorous derivation of nonlinear plate theory and geometric rigidity. C. R. Math. Acad. Sci. Paris **334**(2), 173–178 (2002)
25. N. Fusco, G. Moscariello, C. Sbordone, The limit of $W^{1,1}$ homeomorphisms with finite distortion. Calc. Var. Part. Differ. Equat. **33**(3), 377–390 (2008)
26. F.P. Gardiner, *Teichmüller Theory and Quadratic Differentials*. Pure and Applied Mathematics (New York) (Wiley, New York, 1987). A Wiley-Interscience Publication
27. F.W. Gehring, The definitions and exceptional sets for quasiconformal mappings. Ann. Acad. Sci. Fenn. Ser. A I No. **281**, 28 (1960)
28. F.W. Gehring, Rings and quasiconformal mappings in space. Trans. Am. Math. Soc. **103**, 353–393 (1962)
29. F.W. Gehring, Quasiconformal mappings in Euclidean spaces, in *Handbook of Complex Analysis: Geometric Function Theory*, vol. 2 (Elsevier, Amsterdam, 2005), pp. 1–29
30. F.W. Gehring, J. Väisälä, The coefficients of quasiconformality of domains in space. Acta Math. **114**, 1–70 (1965)
31. M. Giaquinta, *Multiple Integrals in the Calculus of Variations and Nonlinear Elliptic Systems*. Annals of Mathematics Studies, vol. 105 (Princeton University Press, Princeton, 1983)
32. H. Grötzsch, Über die Verzerrung bei schlichten nichtkonformen Abbildungen und über eine damit zusammenhängende Erweiterung des Picardschen Satzes. Berichte Leipzig **80**, 503–507 (1928)
33. R.S. Hamilton, Extremal quasiconformal mappings with prescribed boundary values. Trans. Am. Math. Soc. **138**, 399–406 (1969)
34. J. Heinonen, P. Koskela, Quasiconformal maps in metric spaces with controlled geometry. Acta Math. **181**(1), 1–61 (1998)
35. S. Hencl, P. Koskela, Regularity of the inverse of a planar Sobolev homeomorphism. Arch. Ration. Mech. Anal. **180**(1), 75–95 (2006)
36. S. Hencl, P. Koskela, J. Onninen, A note on extremal mappings of finite distortion. Math. Res. Lett. **12**(2–3), 231–237 (2005)
37. S. Hencl, P. Koskela, J. Onninen, Homeomorphisms of bounded variation. Arch. Ration. Mech. Anal. **186**(3), 351–360 (2007)
38. T. Iwaniec, G. Martin, Quasiregular mappings in even dimensions. Acta Math. **170**(1), 29–81 (1993)
39. T. Iwaniec, G. Martin, *Geometric Function Theory and Non-Linear Analysis*. Oxford Mathematical Monographs (The Clarendon Press/Oxford University Press, New York, 2001)
40. T. Iwaniec, V. Šverák, On mappings with integrable dilatation. Proc. Am. Math. Soc. **118**(1), 181–188 (1993)
41. T. Iwaniec, P. Koskela, J. Onninen, Mappings of finite distortion: monotonicity and continuity. Invent. Math. **144**(3), 507–531 (2001)
42. T. Iwaniec, P. Koskela, J. Onninen, Mappings of finite distortion: compactness. Ann. Acad. Sci. Fenn. Math. **27**(2), 391–417 (2002)

43. T. Iwaniec, L.V. Kovalev, J. Onninen, Lipschitz regularity for inner-variational equations. Duke Math. J. **162**(4), 643672 (2013)
44. R. Jensen,   Uniqueness of Lipschitz extensions: minimizing the sup norm of the gradient. Arch. Ration. Mech. Anal. **123**(1), 51–74 (1993)
45. J. Jost, in *Compact Riemann Surfaces*, 3rd edn. Universitext (Springer, Berlin, 2006), xviii+277 pp.
46. N. Katzourakis, *Extremal Infinity-Quasiconformal Immersions*. Preprint (2012)
47. N.I. Katzourakis, L^∞ variational problems for maps and the Aronsson PDE system. J. Differ. Equat. **253**(7), 2123–2139 (2012)
48. J. Liouville, J. Math. Pure Appl. **15**, 103 (1850)
49. E.J. McShane, Extension of range of functions. Bull. Am. Math. Soc. **40**(12), 837–842 (1934)
50. C. Morrey, Quasiconvexity and the lower semioontinuity of multiple integrals. Pac. J. Math. **2**, 25–53 (1952)
51. C. Morrey, *Multiple Integrals in the Calculus of Variations* (Springer, Berlin, 1966)
52. A. Naor, S. Sheffield, Absolutely minimal Lipschitz extension of tree-valued mappings. Math. Ann. **354**(3), 1049–1078 (2012)
53. Y.-L. Ou, T. Troutman, F. Wilhelm, Infinity-harmonic maps and morphisms. Differ. Geom. Appl. **30**(2), 164–178 (2012)
54. E. Reich,   Extremal quasiconformal mappings of the disk, in *Handbook of Complex Analysis: Geometric Function Theory*, vol. 1 (North-Holland, Amsterdam, 2002), pp. 75–136
55. E. Reich, K. Strebel, Extremal plane quasiconformal mappings with given boundary values. Bull. Am. Math. Soc. **79**, 488–490 (1973)
56. E. Reich, K. Strebel,   Extremal quasiconformal mappings with given boundary values, in *Contributions to Analysis (A Collection of Papers Dedicated to Lipman Bers)* (Academic, New York, 1974), pp. 375–391
57. J.G. Rešetnjak, Liouville's conformal mapping theorem under minimal regularity hypotheses. Sibirsk. Mat. Ž. **8**, 835–840 (1967)
58. Y.G. Reshetnyak, *Space Mappings with Bounded Distortion*. Translations of Mathematical Monographs, vol. 73 (American Mathematical Society, Providence, 1989). Translated from the Russian by H. H. McFaden
59. J. Sarvas,   Ahlfors' trivial deformations and Liouville's theorem in $\mathbf{R}^n$, in *Complex Analysis Joensuu 1978 (Proc. Colloq., Univ. Joensuu, Joensuu, 1978)*. Lecture Notes in Mathematics, vol. 747 (Springer, Berlin, 1979), pp. 343–348
60. V.I. Semenov,   Necessary conditions in extremal problems for spatial quasiconformal mappings. Sibirsk. Mat. Zh. **21**, 5 (1980)
61. V.I. Semenov,   On sufficient conditions for extremal quasiconformal mappings in space. Sibirsk. Mat. Zh. **22**, 3 (1981)
62. V.I. Semënov, S.I. Sheenko,   Some extremal problems in the theory of quasiconformal mappings. Sibirsk. Mat. Zh. **31**, 1 (1990)
63. S. Sheffield, C.K. Smart, *Vector-Valued Optimal Lipschitz Extensions*. Preprint (2010)
64. K. Strebel,   *Quadratic Differentials*. Ergebnisse der Mathematik und ihrer Grenzgebiete (3), vol. 5 [Results in Mathematics and Related Areas (3)] (Springer, Berlin, 1984)
65. K. Strebel,   Extremal quasiconformal mappings. Results Math. **10**(1–2), 168–210 (1986)
66. J. Väisälä,   Two new characterizations for quasiconformality. Ann. Acad. Sci. Fenn. Ser. A I No. **362**, 12 (1965)
67. J. Väisälä,   *Lectures on n-Dimensional Quasiconformal Mappings*. Lecture Notes in Mathematics, vol. 229 (Springer, Berlin, 1971)
68. E. Villamor, J.J. Manfredi,   An extension of Reshetnyak's theorem. Indiana Univ. Math. J. **47**(3), 1131–1145 (1998)
69. H. Whitney,   Analytic extensions of differentiable functions defined in closed sets. Trans. Am. Math. Soc. **36**(1), 63–89 (1934)

Curvature Measures, Isoperimetric Type Inequalities and Fully Nonlinear PDEs

Pengfei Guan

Abstract The notes consider two special fully nonlinear partial differential equations arising from geometric problems, one is of elliptic type and another is of parabolic type. The elliptic equation is associated to the problem of prescribing curvature measures, while an inverse mean curvature type of parabolic equation is introduced to prove the isoperimetric type inequalities for quermassintegrals of k-convex starshaped domains.

The material in the notes is compiled from the lectures given in the CIME Summer School in Cetraro, 2012. It treats some nonlinear elliptic and parabolic partial differential equations arising from geometric problems of hypersurfaces in $\mathbb{R}^{n+1}$. A curvature type of elliptic equation is used to solve the problem of prescribing curvature measures, which is a Minkowski type problem. An inverse mean curvature type of parabolic equation is employed for the proof of isoperimetric type inequalities for quermassintegrals of k-convex starshaped domains. Both types of equations are fully nonlinear, they belong to the category of general geometric fully nonlinear PDE.

The emphasis of the notes is the a priori estimates, which is the key in the theory of fully nonlinear PDE. These estimates are intend to be self-contained here, with minimal assumptions on basic knowledge in PDE and geometry, namely the standard maximum principles for linear elliptic and parabolic equations, the elementary formulas of Gauss, Codazzi and Weingarten for hypersurfaces in $\mathbb{R}^{n+1}$, and the curvature commutator identities. Two theorems we would use without proof for higher regularity are: the Evans-Krylov Theorem [11, 31] for uniformly fully nonlinear elliptic equations and the Krylov Theorem [31] for uniformly parabolic

P. Guan (✉)
Department of Mathematics, McGill University, Montreal, QC H3A 2K6, Canada
e-mail: guan@math.mcgill.ca

L. Capogna et al., *Fully Nonlinear PDEs in Real and Complex Geometry and Optics*,
Lecture Notes in Mathematics 2087, DOI 10.1007/978-3-319-00942-1_2,
© Springer International Publishing Switzerland 2014

fully nonlinear PDE, since the proofs of these deep results would take considerable space.

The topics dealt in this notes are special samples of geometric nonlinear PDE. It is our hope they can serve as an introduction to the general theory of geometric analysis.

The notes are organized as follows. The curvature measures are introduced through the Steiner formula in differential geometric setting in Sect. 1, where the Steiner formula and the Minkowski identity are proved. As the geometric objects and the associated differential equations are involved the elementary symmetric functions, some important properties of these functions are collected in Sect. 2 with proofs, except the theory of hyperbolic polynomials of Garding which is put in the Appendix. Section 3 deals with the problem of prescribing curvature measures. A k-curvature fully nonlinear elliptic equation is set up there together with the a priori estimates of the solutions of the equation. Section 4 is devoted to the proof of the isoperimetric inequalities for quermassintegrals of k-convex star shaped domains, via parabolic approach. Again, the main part is the a priori estimates for the solutions of the corresponding parabolic equation. The literature comments appear at the end of the notes.

1   The Steiner Formula and Curvature Measures

Suppose Ω is a domain in $\mathbb{R}^{n+1}$, for each $x \in \mathbb{R}^{n+1}$, denote $p(\Omega, x)$ to be the set of the nearest points in Ω to x. Given any Borel set $\beta \in \mathfrak{B}(\mathbb{R}^{n+1})$, $\forall s > 0$, consider

$$A_s(\Omega, \beta) := \{x \in \mathbb{R}^{n+1} | 0 < d(\Omega, x) \le s \text{ and } p(\Omega, x) \in \beta\}$$

which is the set of all points $x \in \mathbb{R}^{n+1}$ for which the distance $d(\Omega, x) \le s$ and for which the nearest point $p(\Omega, x)$ belongs to β. If $\partial\Omega$ is smooth and β is open, for $s > 0$ small, one may write

$$A_s(\Omega, \beta) = \{X + t\nu(X) \quad |X \in \beta \cap M, 0 \le t \le s,\}$$

where $\nu(X)$ is the outer normal of M at X.

We assume the boundary of Ω, $M = \partial\Omega$, is C^2 (or smoother). Let

$$\kappa(X) = (\kappa_1(X), \cdots, \kappa_n(X))$$

be the principal curvatures of $X \in M$. To calculate the volume of $A_s(\Omega, \beta)$, pick any local orthonormal frame of M, so that the second fundamental form $(W_{ij}(X))$ of M at X is diagonal. As $(X + t\nu(X))_i = (1 + tW_{ii})X_i$, and $\nu(X)$ is orthogonal to X_i, the volume element at $X + t\nu(X)$ is simply

$$dV = (\prod_{i=1}^{n}(1 + tW_{ii}))d\mu_M dt = \sum_{i=0}^{n} \sigma_i(\kappa(X))t^i d\mu_M dt,$$

where $\sigma_i(\kappa)$ is the i-th elementary symmetric function of κ (see Definition (6)), and where $d\mu_g$ is the volume element with respect to the induced metric g of M in $\mathbb{R}^{n+1}$. Therefore,

$$V(A_s(\Omega,\beta)) = \int_0^s \int_{\beta \cap M} \sum_{i=0}^n \sigma_i(\kappa(X)) t^i d\mu_M dt = \sum_{i=0}^n (\int_{\beta \cap M} \sigma_i(\kappa(X)) d\mu_M) \frac{s^{i+1}}{i+1}.$$

Set

$$\mathcal{C}_m(\Omega) = \sigma_{n-m}(\kappa) d\mu_M, \quad m = 0, 1, \cdots, n. \tag{1}$$

We have proved the Steiner formula,

$$V(A_s(\Omega,\beta)) = \sum_{m=0}^n \frac{s^{n+1-m}}{n+1-m} \mathcal{C}_m(\Omega,\beta), \tag{2}$$

for $\beta \in \mathfrak{B}(\mathbb{R}^{n+1})$ and $s > 0$.

In the context of classical convex geometry, the coefficients $\mathcal{C}_0(\Omega,\cdot), \cdots, \mathcal{C}_n(\Omega,\cdot)$ in (2) are called curvature measures of the convex body Ω. Formula (1) indicates that $\mathcal{C}_m(\Omega,\cdot)$ is well defined if $\partial\Omega$ is C^2 without convexity assumption. In general, $\mathcal{C}_m(\Omega)$ is a signed measure. The positivity of $\mathcal{C}_m(\Omega)$ for $0 \le m \le k$ is related to the notion of k-convexity (Definition 3.1).

The global quantities

$$V_{n-m}(\Omega) = C_{n,k} \int_M \sigma_m(\kappa) d\mu_M, \quad m = 0, 1, \cdots, n, \tag{3}$$

where $C_{n,k} = \frac{\sigma_k(1,\cdots,1)}{\sigma_{k-1}(1,\cdots,1)}$, are called the quermassintegrals of Ω in convex geometry, if Ω is convex. Again, we note that these quantities are well defined for general C^2 domain Ω without convexity condition.

It is clear that the curvature measures capture the geometry of M.

1. What are the relations between quermassintegrals?
2. How much information can we extract from the curvature measures?

These are the main questions we want to deal with in this notes. The first question has satisfactory answer when Ω is convex, which corresponds to the classical Alexandrov-Fenchel inequalities. Generalization of these inequalities to non-convex domains has gained much interest recently, but remains largely unsettled. We will focus on a class of non-convex star-shaped domains, where a clean result can be established. The second question can be answered in terms of the Minkowski type problem, the problem of prescribing curvature measures. It turns out there is an affirmative answer if we restrict ourselves to the class of non-convex star-shaped domains.

There is a different expression for $V_{n-m}(\Omega)$ involving the support function $u(X) = \langle X, \nu(X) \rangle$. The Minkowski identity states that $\forall k \geq 1$,

$$\int_M u\sigma_k(\kappa)d\mu_M = C_{n,k}\int_M \sigma_{k-1}(\kappa)d\mu_M, \tag{4}$$

By the Divergent theorem,

$$V_{n+1}(\Omega) = \frac{1}{n+1}\int_M u\,d\mu_M.$$

From (4), we may define

$$V_{(n+1)-k}(\Omega) = \int_M u\sigma_k(\kappa)d\mu_M, \tag{5}$$

for $k = 0, \cdots, n$. $V_{n+1}(\Omega)$ is multiple of the volume of Ω by a dimensional constant, $V_n(\Omega)$ is a multiple of the surface area of $\partial\Omega$ by another dimensional constant. In convex geometry, u is called the support function of Ω.

The Minkowski identity (4) can be verified using the fact that σ_k has divergent free structure (Lemma 2.1). Again, pick a local orthonormal frame on M, let $h = (W_{ij})$ be the second fundamental from and let $g^{-1}h = (h^i_j)$ be the Weingarten tensor. We compute

$$(\frac{|X|^2}{2})_{ij} = X_i X_j + X_{ij} = \delta_{ij} - \langle X, \nu(X)\rangle W_{ij} = \delta_{ij} - uW_{ij}.$$

Contracting with $\sigma_k^{ij} = \frac{\partial\sigma_k}{\partial h^i_j}(g^{-1}h)$ and integrating over M

$$\int_M \sigma_k^{ij}(\frac{|X|^2}{2})_{ij} = \int_M (\sum_i \sigma_k^{ij}\delta_{ij} - u\sigma_k^{ij}W_{ij}).$$

As

$$\sigma_k^{ij}\delta_{ij} = (n-k+1)\sigma_{k-1}, \quad \sigma_k^{ij}W_{ij} = k\sigma_k,$$

and by (8), we get

$$0 = (n-k+1)\int_M \sigma_{k-1}(g^{-1}h) - k\int_M u\sigma_k(g^{-1}h).$$

This is exactly the identity (4).

The Minkowski addition of two sets $\Omega_1, \Omega_2 \subset \mathbb{R}^{n+1}$ is defined as

$$\Omega_1 + \Omega_2 = \{z = x + y | x \in \Omega_1, y \in \Omega_2\}.$$

The Minkowski addition is one of the basic operation in convex geometry. For general domain Ω, when $0 \le s$ small, one may define

$$\Omega_s = \{z = x + y \,|\, x \in \Omega, y \in B_s\},$$

where B_s is the ball centered at the origin with radius s.

$$\Omega_s = \{X + t\nu(X) \quad |X \in \Omega, 0 \le t \le s.\}$$

If $M = \partial\Omega$ is smooth, the boundary $\partial\Omega_s = M_s$ is also smooth and can be written as

$$M_s = \{X + s\nu(X) \quad |X \in M.\}$$

Moreover, the normal of M_s at $X_s = X + s\nu(X)$ is the same as $\nu(X)$ for each $X \in M$. The support function of Ω_s is $u_s(X^s) = u(X) + s$. For any local orthonormal frame $e_1, \cdots, e_n$ on M such that $h = (W_{ij})$ is diagonal at the point, one may calculate the induced metric g_s on M^s

$$g_s = \sum_{i=1}^{n} (1 + h_i^i)^2 e_i \otimes e_i,$$

and the area element of M^s

$$d\mu_{M_s} = \det(I + sg^{-1}h)d\mu_M.$$

By the Minkowski identity, the volume of Ω_s can be computed as

$$V(\Omega_s) = \frac{1}{n+1} \int_M u_s \det(I + sg^{-1}h)d\mu_M$$

$$= \frac{1}{n+1} \int_M \sum_{i=0}^{n} (u + s)s^i \sigma_i(g^{-1}h)d\mu_M$$

$$= \frac{1}{n+1} \int_M \sum_{i=0}^{n} (us^i \sigma_i(g^{-1}h) + s^{i+1})\sigma_i(g^{-1}h)d\mu_M$$

$$= \frac{1}{n+1} \sum_{i=0}^{n} \frac{n+1}{i+1} s^{i+1} \int_M \sigma_i(g^{-1}h)d\mu_M + \frac{1}{n+1} \int_M u d\mu_M$$

$$= \sum_{i=0}^{n+1} c_{n+1}^i t^{n+1-i} V_i(\Omega),$$

2 Some Properties of Elementary Symmetric Functions

The elementary symmetric functions appear naturally in the geometric quantities in the previous section. In order to carry on analysis, we need to understand properties of the elementary symmetric functions.

For $1 \leq k \leq n$, and $\lambda = (\lambda_1, \ldots, \lambda_n) \in \mathbb{R}^n$, the k-th elementary symmetric function is defined as

$$\sigma_k(\lambda) = \sum \lambda_{i_1} \ldots \lambda_{i_k}, \tag{6}$$

where the sum is taken over all strictly increasing sequences $i_1, \ldots, i_k$ of the indices from the set $\{1, \ldots, n\}$. The definition can be extended to symmetric matrices. Denote $\lambda(W) = (\lambda_1(W), \ldots, \lambda_n(W))$ to be the eigenvalues of the symmetric matrix W, set

$$\sigma_k(W) = \sigma_k(\lambda(W)).$$

It is convenient to set

$$\sigma_0(W) = 1, \quad \sigma_k(W) = 0, \quad \text{for } k > n.$$

It follows directly from the definition that, for any $n \times n$ symmetric matrix W, and $\forall t \in \mathbb{R}$,

$$\sigma_n(I + tW) = \det(I + tW) = \sum_{i=0}^{n} \sigma_i(W)t^i. \tag{7}$$

Conversely, (7) can also be used to define $\sigma_k(W)$, $\forall k = 0, \cdots, n$.

An important property of σ_k is the divergent free structure. Suppose M is a general Riemannian manifold of dimension n, W is a symmetric tensor on M. We call W is Codazzi if $DW = 0$. This property is equivalent to say that, for any local orthonormal frame $(e_1, \cdots, e_n)$ on M, write $W = (w_{ij})$, then $w_{ij,l} = \nabla_{e_l} w_{ij}$ is symmetric with respect to i, j, l. Some classical examples are

1. Second fundamental form h of any hypersurface in space form $N(c)$ with constant sectional curvature c, this follows from the Codazzi equation;
2. $W = \overline{\nabla}^2 v + cv, \forall v \in C^3(N(c))$.

Throughout the rest of the notes, we will use Einstein summation convention, unless it is otherwise indicated.

Below is the statement of divergent free structure of σ_k.

Lemma 2.1. *Suppose $e_1, \cdots, e_n$ is a local orthonormal frame on M, $W = (w_{ij})$ is a Codazzi tensor on M, then for each i,*

$$\sum_{j=1}^{n} \left(\frac{\partial \sigma_k}{\partial w_{ij}}\right)_j (W) = 0. \tag{8}$$

Proof. We first verify (8) for $k = n$. Denote C^{il} to be the cofactor of W, i.e.,

$$\frac{\partial \sigma_n}{\partial w_{il}} = C^{il}, \quad C^{il} w_{lj} = \det(W)\delta_j^i.$$

Differentiate above identity in e_m direction and contract with C^{jm},

$$C^{jm} C_m^{il} w_{lj} + C^{il} w_{lj,m} C^{jm} = \delta_j^i (\det(W))_m C^{jm}.$$

If $\det(W) \neq 0$ at the point, we get

$$C_m^{im} = C^{pq} C^{im} w_{pq,m} - C^{il} C^{jm} w_{lj,m} = C^{pq} C^{im} w_{pq,m} - C^{il} C^{jm} w_{jm,l} = 0.$$

If $\det(W) = 0$ at the point, we may approximate W by Codazzi tensor $\tilde{W} = W + tg$ where g is the metric tensor on M such that $\det(\tilde{W}) \neq 0$ for t small. Equation (8) is verified for the case $k = n$.

Observe that, for $t \in \mathbb{R}$,

$$\sigma_n(\tilde{W}) = \sum_{m=0}^{n} t^m \sigma_{n-m}(W).$$

Apply (8) for the case $k = n$,

$$\sum_{m=0}^{n} t^m \sum_j (\frac{\partial \sigma_{n-m}}{\partial w_{ij}}(W))_j = 0.$$

Since it is true for all $t \in \mathbb{R}$, we must have $\forall m$,

$$\sum_j (\frac{\partial \sigma_{n-m}}{\partial w_{ij}}(W))_j = 0.$$

$\square$

The following gives explicit algebraic formulas for $\sigma_k(W)$.

Proposition 2.2. *If $W = (W_{ij})$ is an $n \times n$ symmetric matrix, let $F(W) = \sigma_k(W)$ for $1 \leq k \leq n$. Then the following relations hold.*

$$\sigma_k(W) = \frac{1}{k!} \sum_{\substack{i_1,\ldots,i_k=1 \\ j_1,\ldots,j_k=1}}^{n} \delta(i_1,\ldots,i_k; j_1,\ldots,j_k) W_{i_1 j_1} \cdots W_{i_k j_k},$$

$$F^{\alpha\beta} := \frac{\partial F}{\partial W_{\alpha\beta}}(W)$$

$$= \frac{1}{(k-1)!} \sum_{\substack{i_1,\ldots,i_{k-1}=1 \\ j_1,\ldots,j_{k-1}=1}}^{n} \delta(\alpha,i_1,\ldots,i_{k-1};\beta,j_1,\ldots,j_{k-1})W_{i_1 j_1}\cdots W_{i_{k-1}j_{k-1}}$$

$$F^{ij,rs} := \frac{\partial^2 F}{\partial W_{ij}\partial W_{rs}}(W)$$

$$= \frac{1}{(k-2)!} \sum_{\substack{i_1,\ldots,i_{k-2}=1 \\ j_1,\ldots,j_{k-2}=1}}^{n} \delta(i,r,i_1,\ldots,i_{k-2};j,s,j_1,\ldots,j_{k-2})W_{i_1 j_1}\cdots W_{i_{k-2}j_{k-2}},$$

where the Kronecker symbol $\delta(I;J)$ for indices $I = (i_1,\ldots,i_m)$ and $J = (j_1,\ldots,j_m)$ is defined as

$$\delta(I;J) = \begin{cases} 1, & \text{if } I \text{ is an even permutation of } J; \\ -1, & \text{if } I \text{ is an odd permutation of } J; \\ 0, & \text{otherwise.} \end{cases}$$

$$\sum_{i,j,m} \sigma_k^{ij}(W)W_{im}W_{mj} = \sigma_1(W)\sigma_k(W) - (k+1)\sigma_{k+1}(W).$$

Proof. The first identity follows from (7) by equalized the coefficient in front of t^k. The second and third identities follow from the first identity. Notice that all the identity are invariant under orthornormal transformation. In particular, we may assume W is diagonal in the last identity. For $\lambda \in \mathbb{R}^n$, for any fixed $i \in \{1,\cdots,n\}$, denote $(\lambda|i) \in \mathbb{R}^n$ with i-th component of λ replaced by 0. Differentiation of (6) yields

$$\frac{\partial \sigma_k(\lambda)}{\partial \lambda_i} = \sigma_{k-1}(\lambda|i). \tag{9}$$

Again it can read off from (6),

$$\sigma_k(\lambda) = \sigma_k(\lambda|i) + \lambda_i \sigma_{k-1}(\lambda|i). \tag{10}$$

Thus,

$$\lambda_i \sigma_k(\lambda|i) = \lambda_i(\sigma_k(\lambda) - \lambda_i \sigma_{k-1}(\lambda|i) = \lambda_i \sigma_k(\lambda) - \lambda_i^2 \sigma_{k-1}(\lambda|i).$$

Using homogeneity of σ_{k+1}, the last identity in the proposition follows from the above by summing up over i. □

Definition 2.3. For $1 \leq k \leq n$, let Γ_k is a cone in $\mathbb{R}^n$ determined by

$$\Gamma_k = \{\lambda \in \mathbb{R}^n : \quad \sigma_1(\lambda) > 0, \ldots, \sigma_k(\lambda) > 0\}.$$

A $n \times n$ symmetric matrix W is called belong to Γ_k is $\lambda(W) \in \Gamma_k$.

Let $W^1, \cdots, W^n$ be $n \times n$ symmetric matrices, define $\sigma_n(W^1, \ldots, W^n)$ to be the coefficient in front of the factor $t_1 \cdots t_n$ of the polynomial $\det(t_1 W^1 + \cdots + t_n W^n)$. It is called the mixed determinant of $W^1, \cdots, W^n$. In general, for $1 \leq k \leq n$, we define $\sigma_k(W^1, \ldots, W^k) = \binom{n}{k}\sigma_n(W^1, \ldots, W^k, I, \cdots, I)$, where the identity matrix I appears $(n-k)$ times. $\sigma_k(W^1, \ldots, W^k)$ is called the complete polarization of the symmetric function σ_k.

The following Garding inequality plays important role in geometric PDE.

Lemma 2.4. Γ_k is a convex cone. $\forall W^i \in \Gamma_k, i = 1, \ldots, k$,

$$\sigma_k^2(W^1, W^2, W^3, \cdots, W^k) \geq \sigma_k(W^1, W^1, W^3, \cdots, W^k)\sigma_k(W^2, W^2, W^3, \cdots, W^k),$$
$$(11)$$

equality hold if and only if W^1 and W^2 are proportional. And

$$\sigma_k(W^1, \cdots, W^k) \geq \sigma_k^{\frac{1}{k}}(W^1, \cdots, W^1) \cdots \sigma_k^{\frac{1}{k}}(W^k, \cdots, W^k),$$
$$(12)$$

the equality holds if and only if W^i, W^j are pairwise proportional.

Lemma 2.4 is a special case of Garding's theory of hyperbolic polynomials, which can be found in Appendix. The convexity of Γ_k follows from Proposition 5.2, (11) and (12) follow from Corollary 5.4 and Proposition 5.6 in Appendix.

Inequality (11) yields the Newton-MacLaurin inequality.

Lemma 2.5. *For $W \in \Gamma_k$,*

$$(n - k + 1)(k + 1)\sigma_{k-1}(W)\sigma_{k+1}(W) \leq k(n - k)\sigma_k^2(W),$$
$$(13)$$

and

$$\sigma_{k+1}(W) \leq c_{n,k}\sigma_k^{\frac{k+1}{k}}(W),$$
$$(14)$$

where $c_{n,k} = \frac{\sigma_{k+1}(I)}{\sigma_k}^{\frac{k+1}{k}}(I)$. The equality holds if and only if $W = cI$ for some $c > 0$.

Proof. If $\sigma_{k+1}(W) \leq 0$, as $W \in \Gamma_k$, (13) is trivial. We may assume $\sigma_{k+1}(W) > 0$, so $W \in \Gamma_{k+1}$. Replace k by $k + 1$ in (11), and set $W^1 = I$, $W^2 = \cdots = W^{k+1} = W \in \Gamma_k$, (13) follows from (11). The similar argument yields (14) using (12). $\quad\square$

We remark that the Newton-MacLaurin inequality is valid for general symmetric matrix W (e.g., [28]).

The following lemma establish connection of σ_k with the ellipticity of Hessian and curvature equations.

Lemma 2.6. *Let $F = \sigma_k$, then the matrix $(\frac{\partial F}{\partial W_{ij}})$ is positive definite for $W \in \Gamma_k$. where W_{ij} are the entries of W. If $W \in \Gamma_k$, then $(W|i) \in \Gamma_{k-1}, \forall k = 0, 1, \cdots, n$, $i = 1, 2, \cdots, n$, where $(W|i)$ is the matrix with i-th column and i-th row deleted. Furthermore, if $W \in \Gamma_k$ and $\|W\| = \sqrt{\sum_{i,j} w_{ij}^2} \leq R$ for some $R > 0$, then there is $c_{n,k} > 0$ depending only on n, k, such that*

$$\frac{\sigma_k(W)}{R(1 + c_{n,k}\sigma_{k-1}^{\frac{1}{k-1}}(I))} I \leq (\frac{\partial F}{\partial W_{ij}}) \leq R^{k-1}\sigma_{k-1}(I)I. \tag{15}$$

Proof. Fix $W \in \Gamma_k$, for any positive definite matrix $A = (a_{ij})$, by Lemma 2.4,

$$0 < \sigma_k(W, \cdots, W, A) = \sum_{ij} \frac{\partial F}{\partial w_{ij}}(W)a_{ij}.$$

This implies the positivity of $(\frac{\partial F}{\partial W_{ij}})$. By Proposition 2.2 and the positivity of $(\frac{\partial F}{\partial W_{ij}})$, for each $l \leq k$, $W \in \Gamma_k$, and for any $i \in \{1, \cdots, n\}$,

$$0 < \frac{\partial \sigma_l}{\partial W_{ii}} = \sigma_{l-1}(W|i).$$

This yields $(W|i) \in \Gamma_{k-1}$.

To show (15), we only need to control $\frac{\partial \sigma_l}{\partial \lambda_i} = \sigma_{k-1}(\lambda|i)$, where $\lambda_i, i = 1, \cdots, n$ are the eigenvalues of W. By the assumption, and (14)

$$s \leq \sigma_k(W) = \sigma_k(\lambda|i) + \lambda_i\sigma_{k-1}(\lambda|i)$$

$$\leq \sigma_{k-1}(\lambda|i)(\lambda_i + c_{n,k}\sigma_{k-1}^{\frac{1}{k-1}}(\lambda|i))$$

$$\leq R(1 + c_{n,k}\sigma_{k-1}^{\frac{1}{k-1}}(I))\sigma_{k-1}(\lambda|i).$$

this gives the lower bound in (15). The upper bound for $\sigma_{k-1}(\lambda|i)$ is trivial. $\square$

We now switch to the quotient of elementary symmetric functions. Some of the concave properties of them will be used in crucial way in the a priori estimates in the rest of the sections.

Lemma 2.7. *For $0 \leq l < k \leq n$, let $F = (\frac{\sigma_k}{\sigma_l})^{\frac{1}{k-1}}$, then $(\frac{\partial F}{\partial w_{ij}})$ is positive definite for $W = (w_{ij}) \in \Gamma_k$. If $l = k - 1$, if $W \in \Gamma_k$ and $\|W\| = \sqrt{\sum_{i,j} w_{ij}^2} \leq R$ for some $R > 0$, then there is $c_{n,k} > 0$ depending only on n, k, such that*

$$\frac{F(W)}{R(1 + c_{n,k}\sigma_{k-1}^{\frac{1}{k-1}}(I))} I \leq (\frac{\partial F}{\partial w_{ij}}) \leq (n - k + 1)I. \tag{16}$$

Moreover, the function F is concave in Γ_{k-1}.

Proof. To simplify notation, define

$$Q_m = \frac{\sigma_m}{\sigma_{m-1}}.$$

For any $l < k$,

$$\frac{\sigma_k}{\sigma_l} = \prod_{j=1}^{k-l} Q_{l+j}. \tag{17}$$

As $Q_{l+j} > 0$ for $j = 1, \cdots, k - l$, for the first statement in lemma, we only need to check the positivity of $(\frac{\partial Q_{m+1}(W)}{\partial w_{ij}})$ for $W = (w_{ij}) \in \Gamma_k$ and for $m = l, \cdots, k - 1$. By product rule,

$$\frac{\partial Q_{m+1}(W)}{\partial w_{ij}} = \frac{\sigma_m(W)\frac{\partial \sigma_{m+1}(W)}{\partial w_{ij}} - \sigma_{m+1}(W)\frac{\partial \sigma_m(W)}{\partial w_{ij}}}{\sigma_m^2(W)}.$$

By Proposition 2.2, the positivity of $(\frac{\partial \sigma_j(W)}{\partial w_{ij}})$ is invariant under orthonormal transformations, we only need to check the positivity of $\frac{\partial Q_{m+1}(\lambda)}{\partial \lambda_i}$ for $\lambda \in \Gamma_k$, $i \in \{1, \cdots, n\}$ and $m = l, \cdots, k - 1$. Again,

$$\begin{aligned}
\frac{\partial Q_{m+1}(\lambda)}{\partial \lambda_i} &= \frac{\sigma_m(\lambda)\frac{\partial \sigma_{m+1}(\lambda)}{\partial \lambda_i} - \sigma_{m+1}(\lambda)\frac{\partial \sigma_m(\lambda)}{\partial \lambda_i}}{\sigma_m^2(\lambda)} \\
&= \frac{\sigma_m(\lambda)\sigma_m(\lambda|i) - \sigma_{m+1}(\lambda)\sigma_{m-1}(\lambda|i)}{\sigma_m^2(\lambda)} \\
&= \frac{\sigma_m(\lambda|i)\sigma_m(\lambda|i) - \sigma_{m+1}(\lambda|i)\sigma_{m-1}(\lambda|i)}{\sigma_m^2(\lambda)} \\
&\geq \frac{n}{(n-m)(m+1)} \frac{\sigma_m^2(\lambda|i)}{\sigma_m^2(\lambda)} \\
&> 0,
\end{aligned} \tag{18}$$

the Newton-MacLaurine inequality (13) is used in the last step as $(\lambda|i) \in \Gamma_{k-1}$ for each i. In particular, if $m = k - 1$ and $W \in \Gamma_k$, for each i,

$$\begin{aligned}
0 < \frac{\partial Q_k(\lambda)}{\partial \lambda_i} &\leq \sum_i \frac{\partial Q_k(\lambda)}{\partial \lambda_i} \\
&\leq \sum_i \frac{\sigma_{k-1}(\lambda|i)}{\sigma_{k-1}(\lambda)} \\
&= n - k + 1.
\end{aligned}$$

This provides the upper bound in (16). By (14)

$$\frac{\sigma_k(W)}{\sigma_{k-1}(W)} \le c_{n,k-1} \sigma_k^{\frac{1}{k}}(W).$$

For each $i = 1, \cdots, n$,

$$\frac{\sigma_{k-1}(\lambda|i)}{\sigma_{k-1}(\lambda)} = \frac{\sigma_k(\lambda)}{\sigma_{k-1}(\lambda)} \frac{\sigma_{k-1}(\lambda|i)}{\sigma_k(\lambda)}.$$

Now the lower bound in (16) follows from (18) and (15).

Notice that if $f_1 > 0$ and $f_2 > 0$ are two concave function, for any $1 \ge \alpha \ge 0$, $f = f_1^\alpha f_2^{1-\alpha}$ is also concave. Hence, we only need to check the concavity of $\frac{\sigma_{m+1}}{\sigma_m}$ in Γ_{m+1}. In fact, we show $\frac{\sigma_{m+1}}{\sigma_m}$ in Γ_m.

$m = 0$ is trivial. For $m = 1$, there is a useful explicit formula. $\forall \lambda, \lambda \pm \xi \in \Gamma_1$, we have algebraic identity

$$2Q_2(\lambda) - Q_2(\lambda + \xi) - Q_2(\lambda - \xi) = \frac{(\sum_i (\xi_i \sigma_1(\lambda) - \lambda_i \sigma_1(\xi)))^2}{\sigma_1(\lambda)\sigma_1(\lambda + \xi)\sigma_1(\lambda - \xi)}.$$

This yields,

$$\frac{\partial^2 Q_2}{\partial^2 \xi} = -\frac{(\sum_i (\xi_i \sigma_1(\lambda) - \lambda_i \sigma_1(\xi)))^2}{\sigma_1^3(\lambda)}$$

This gives the concavity of $\frac{\sigma_2}{\sigma_1}$ on Γ_1.

For $m > 1$, we use induction. For $\lambda \in \Gamma_m$, for each $i \in \{1, \cdots, n\}$ fixed, by (10) and Corollary 2.6,

$$\lambda_i + Q_m(\lambda|i) = \frac{\sigma_{m+1}(\lambda)}{\sigma_m(\lambda|i)} > 0.$$

Apply the last identity in Proposition 2.2,

$$(m+1)Q_m(\lambda) = \sum_i (\lambda_i - \lambda_i^2 \frac{\sigma_{m-1}(\lambda|i)}{\sigma_m(\lambda)})$$

$$= \sum_i (\lambda_i - \lambda_i^2 \frac{\sigma_{m-1}(\lambda|i)}{\sigma_m(\lambda|i) + \lambda_i \sigma_{m-1}(\lambda|i))})$$

$$= \sum_i (\lambda_i - \frac{\lambda_i^2}{\lambda_i + Q_m(\lambda|i)}).$$

For any $\xi \in \mathbb{R}^n$ with $|\xi| = 1$, set $\lambda_{\epsilon\pm} = \lambda \pm \epsilon\xi$. Take $\epsilon > 0$ small enough such that $\lambda_{\epsilon\pm} \in \Gamma_m$, using the above identity for $\lambda, \lambda_{\epsilon\pm}$, one compute

$$(m+1)(2Q_{m+1}(\lambda) - Q_{m+1}(\lambda_{\epsilon+}) - Q_{m+1}(\lambda_{\epsilon-}))$$

$$= \sum_i \left(\frac{(\lambda_i + \epsilon \xi_i)^2}{Q_m(\lambda_{\epsilon+}|i) + \lambda_i + \epsilon \xi_i} + \frac{(\lambda_i - \epsilon \xi_i)^2}{Q_m(\lambda_{\epsilon-}|i) + \lambda_i - \epsilon \xi_i} \right.$$

$$\left. - \frac{(2\lambda_i)^2}{Q_m(\lambda_{\epsilon+}|i) + Q_m(\lambda_{\epsilon-}|i) + 2\lambda_i} \right)$$

$$+ \sum_i \left(\frac{(2\lambda_i)^2}{Q_m(\lambda_{\epsilon+}|i) + Q_m(\lambda_{\epsilon-}|i) + 2\lambda_i} - \frac{2\lambda_i^2}{\lambda_i + Q_m(\lambda|i)} \right)$$

$$= \sum_i \frac{((\lambda_i + \epsilon \xi_i)Q_m(\lambda_{\epsilon-}) - (\lambda_i - \epsilon \xi_i)Q_m(\lambda_{\epsilon+}))^2}{(Q_m(\lambda_{\epsilon+}) + \lambda_i + \epsilon \xi_i)(Q_m(\lambda_{\epsilon-}) + \lambda_i - \epsilon \xi_i)(Q_m(\lambda_{\epsilon+}) + Q_m(\lambda_{\epsilon-}) + \epsilon \lambda_i)}$$

$$- 2 \sum_i \lambda_i^2 \frac{Q_m(\lambda_{\epsilon+}|i) + Q_m(\lambda_{\epsilon-}|i) - 2Q_m(\lambda)}{(Q_m(\lambda_{\epsilon+}|i) + Q_m(\lambda_{\epsilon-}|i) + 2\lambda_i)(\lambda_i + Q_m(\lambda|i))}$$

Thus,

$$-\frac{\partial^2 Q_{m+1}}{\partial^2 \xi} = \lim_{\epsilon \to 0} \frac{2Q_{m+1}(\lambda) - Q_{m+1}(\lambda_{\epsilon+}) - Q_{m+1}(\lambda_{\epsilon-})}{\epsilon^2}$$

$$\geq \lim_{\epsilon \to 0} -2 \sum_i \lambda_i^2 \frac{Q_m(\lambda_{\epsilon+}|i) + Q_m(\lambda_{\epsilon-}|i) - 2Q_m(\lambda)}{\epsilon^2 (Q_m(\lambda_{\epsilon+}|i) + Q_m(\lambda_{\epsilon-}|i) + 2\lambda_i)(\lambda_i + Q_m(\lambda|i))}$$

$$= -\sum_i \frac{\lambda_i^2 (\frac{\partial^2 Q_m}{\partial \xi^2})(\lambda|i)}{(m+1)(Q_m(\lambda|i) + \lambda_i)^2}.$$

As $(\lambda|i) \in \Gamma_{m-1}$, by induction hypothesis, $\frac{\partial^2 Q_m}{\partial \xi^2}(\lambda|i) \leq 0$. $\qquad\qquad \square$

The following lemma will play key role for the problem of prescribing curvature measures.

Lemma 2.8. *Let* $\alpha = \frac{1}{k-1}$, *if* $W \in \Gamma_k$ *is a symmetric tensor on a Riemannian manifold* M. *For any local orthornormal frame* $\{e_1, \cdots, e_n\}$, *denote* $W_{ij,s} = \nabla_{e_s} W_{ij}$. *Then*

$$(\sigma_k)^{ij,lm} W_{ij,s} W_{lm,s} \leq -\sigma_k \left[\frac{(\sigma_k)_s}{\sigma_k} - \frac{(\sigma_1)_s}{\sigma_1} \right] \left[(\alpha - 1) \frac{(\sigma_k)_s}{\sigma_k} - (\alpha + 1) \frac{(\sigma_1)_s}{\sigma_1} \right]. \quad (19)$$

Proof. By the concavity of $\left(\frac{\sigma_k}{\sigma_1} \right)^{\frac{1}{k-1}} (W)$, we have

$$0 \geq \frac{\partial^2}{\partial W_{ij} \partial W_{lm}} \left(\left(\frac{\sigma_k}{\sigma_1} \right)^{\frac{1}{k-1}} \right) W_{ij,s} W_{lm,s}. \quad (20)$$

Denote $\alpha = \frac{1}{k-1}$. Direct computations yield,

$$0 \geq \frac{\partial^2}{\partial W_{ij} \partial W_{lm}} \left(\frac{\sigma_k}{\sigma_1} \right)^\alpha \cdot W_{ij,s} W_{lm,s}$$

$$= \alpha \left(\frac{\sigma_k}{\sigma_1} \right)^\alpha \left[\frac{(\sigma_k)^{ij,lm}}{\sigma_k} + \frac{(\alpha-1)(\sigma_k)^{ij}(\sigma_k)^{lm}}{\sigma_k^2} \right. \tag{21}$$
$$\left. - \frac{2\alpha(\sigma_k)^{ij}(\sigma_1)^{lm}}{\sigma_k \sigma_1} + \frac{(\alpha+1)(\sigma_1)^{ij}(\sigma_1)^{lm}}{\sigma_1^2} \right] W_{ij,s} W_{lm,s}$$

Equivalently,

$$\frac{(\sigma_k)^{ij,lm} W_{ij,s} W_{lm,s}}{\sigma_k} \leq - \left[\frac{(\alpha-1)(\sigma_k)^{ij}(\sigma_k)^{lm}}{\sigma_k^2} - \frac{2\alpha(\sigma_k)^{ij}(\sigma_1)^{lm}}{\sigma_k \sigma_1} \right.$$
$$\left. + \frac{(\alpha+1)(\sigma_1)^{ij}(\sigma_1)^{lm}}{\sigma_1^2} \right] W_{ij,s} W_{lm,s}$$
$$\tag{22}$$
$$\leq - \left[\frac{(\sigma_k)_s}{\sigma_k} - \frac{(\sigma_1)_s}{\sigma_1} \right] \left[(\alpha - 1)\frac{(\sigma_k)_s}{\sigma_k} - (\alpha + 1)\frac{(\sigma_1)_s}{\sigma_1} \right]$$

$\square$

Note in Lemma 2.8, one may replace σ_k by any positive function F with the property that $(\frac{F}{\sigma_1})^\alpha$ is concave for some $\alpha > 0$. The following is a corollary of Lemma 2.8.

Corollary 2.9. *If* $\frac{(\sigma_1)_s}{\sigma_1} = \frac{(\sigma_k)_s}{\sigma_k} - r$ *for some* r,

$$(\sigma_k)^{ij,lm} W_{ij;s} W_{ij;s} \leq \max \left\{ 2r(\sigma_k)_s - \frac{k}{k-1} r^2 \sigma_k, 0 \right\}. \tag{23}$$

3 Prescribing Curvature Measures

Assume $\Omega \subset \mathbb{R}^{n+1}$ is a bounded star-shaped domain with respect to the origin. We may parametrize $M = \partial\Omega$ over $\mathbb{S}^n$ by positive radial function ρ Due to the parametrization, the prescribe curvature measure problem for this class of domains can be reduced to a curvature type nonlinear partial differential equation of ρ on $\mathbb{S}^n$. We want to establish the existence theorems of prescribing general $(n-k)$-th curvature measure problem with $k > 0$ on bounded C^2 star-shaped domains. When $k = n$, the prescribing curvature measure $\mathcal{C}_0$ is the Alexandrov problem corresponding to a Monge-Ampère type equation on $\mathbb{S}^n$, which won't be treated here.

In order to make the problem in proper PDE setting, we need to impose some geometric condition on $\partial\Omega$.

Definition 3.1. A domain Ω is called k-convex if its principal curvature vector $\kappa(x) = (\kappa_1, \cdots, \kappa_n) \in \Gamma_k$ at every point $x \in \partial\Omega$.

For each star-shaped domain Ω with $M = \partial\Omega$, express M as a radial graph of $\mathbb{S}^n$,

$$R_M : \mathbb{S}^n \longrightarrow M$$
$$z \longmapsto \rho(z)z.$$

From (1) the $(n-k)$-th curvature measure on each Borel set β in $\mathbb{S}^n$ can be defined as

$$C_k(M, \beta) := \int_{R_M(\beta)} \sigma_k(\kappa) d\mu_g.$$

The precise statement of the problem for prescribing $(n-k)$-th curvature measure is: *given a positive function $f \in C^2(\mathbb{S}^n)$, find a closed hypersurface M as a radial graph over $\mathbb{S}^n$, such that $C_{n-k}(M, \beta) = \int_\beta f d\mu$ for every Borel set β in $\mathbb{S}^n$, where $d\mu$ is the standard volume element on $\mathbb{S}^n$.*

For the C^2 graph M on $\mathbb{S}^n$, denote the induced metric to be g and the density function is $\sqrt{\det g}$. Then

$$C_{n-k}(M, \beta) = \int_{R_M(\beta)} \sigma_k d\mu_g = \int_\beta \sigma_k \sqrt{\det g} d\mathbb{S}^n. \tag{24}$$

We now write down the local expressions of the induced metric, support function u, second fundamental form and Weingarten curvatures in terms of positive function ρ and its derivatives $\nabla\rho$, $\nabla^2\rho$. Let $\{e_1, \cdots, e_n\}$ be a local orthonormal frame on $\mathbb{S}^n$, and denote e_{ij} the standard spherical metric with respect to this frame (which is the identity matrix). We use $\overline{\nabla}$ as the gradient operator with respect to standard metric on $\mathbb{S}^n$. To simplify notation, for any function v on $\mathbb{S}^n$, we will write $\overline{\nabla}_{e_i} v = v_i$ as covariant derivative with respect to e_i on $\mathbb{S}^n$ in this subsection, if there is no confusion. From the radial parametrization $X(x) = \rho(x)x$,

$$X_i = \rho_i x + \rho e_i,$$
$$X_{ij} = \rho_{ij} x + \rho_i e_j + \rho_j e_i + \rho(e_i)_j = \rho_{ij} x + \rho_i e_j + \rho_j e_i - \rho e_{ij} x.$$

The following identities can be read off from the above.

$$v = \frac{\rho x - \overline{\nabla}\rho}{\sqrt{\rho^2 + |\overline{\rho}|^2}}$$

$$u = \frac{\rho^2}{\sqrt{\rho^2 + |\overline{\nabla}\rho|^2}}$$

$$g_{ij} = \rho^2 \delta_{ij} + \rho_i \rho_j$$

$$g^{ij} = \frac{1}{\rho^2}(\delta^{ij} - \frac{\rho_i \rho_j}{\rho^2 + |\overline{\nabla}\rho|^2}) \qquad (25)$$

$$h_{ij} = (\sqrt{\rho^2 + |\overline{\nabla}\rho|^2})^{-1}(-\rho\overline{\nabla}_i\overline{\nabla}_j\rho + 2\rho_i\rho_j + \rho^2 e_{ij})$$

$$h^i_j = \frac{1}{\rho^2\sqrt{\rho^2 + |\overline{\nabla}\rho|^2}}(e^{ik} - \frac{\rho_i\rho_k}{\rho^2 + |\overline{\nabla}\rho|^2})(-\rho\overline{\nabla}_k\overline{\nabla}_j\rho + 2\rho_k\rho_j + \rho^2 e_{kj}).$$

From (25),

$$\sqrt{\det g} = \rho^{n-1}\sqrt{\rho^2 + |\overline{\nabla}\rho|^2}.$$

The prescribing $(n - k)$-th curvature measure problem can be deduced to the following curvature equation on $\mathbb{S}^n$:

$$\sigma_k(\kappa_1, \cdots, \kappa_n) = \sigma_k(h^i_j) = \frac{f}{\rho^{n-1}\sqrt{\rho^2 + |\overline{\nabla}\rho|^2}}, \qquad (26)$$

where $f > 0$ is the given function on $\mathbb{S}^n$. A solution of (26) is called admissible if $\kappa(X) \in \Gamma_k$ at each point $X \in M$. We note that any positive C^2 function ρ on $\mathbb{S}^n$ satisfying (26) is automatically an admissible solution. Since the principal curvatures at a maximum point of ρ are positive, solution is admissible at this point. As Γ_k and $\mathbb{S}^n$ are connected, and $\kappa(X)$ varies continuously, the fact of $\sigma_k(\kappa(X)) > 0$ implies solution is admissible at each point of M.

The following is the statement of solvability of the problem of the prescribing curvature measures.

Theorem 3.2. *Let $n \geq 2$ and $1 \leq k \leq n - 1$. Suppose $f \in C^2(\mathbb{S}^n)$ and $f > 0$. Then there exists a unique k-convex star-shaped hypersurface $M \in C^{3,\alpha}$, $\forall \alpha \in (0, 1)$ such that it satisfies (26). Moreover, there is a constant C depending only on $k, n, \|f\|_{C^{1,1}}, \|1/f\|_{C^0}$, and α such that,*

$$\|\rho\|_{C^{3,\alpha}} \leq C. \qquad (27)$$

The rest of the section is devoted to the proof of Theorem 3.2. The main task will be the a priori estimates for solutions of (26). We will use the radial parametrization on $\mathbb{S}^n$ for the estimates up to C^1. Then we will work directly on M for the curvature estimates, which is equivalent to C^2 estimates.

It will be convenient to introduce a new variable $\gamma = \log \rho$. Set

$$\omega := \sqrt{1 + |\overline{\nabla}\gamma|^2}.$$

The unit outward normal and support function can be expressed as $\nu = \frac{1}{\omega}(1, -\gamma_1, \cdots, -\gamma_n)$ and $u = \frac{e^\gamma}{\omega}$ respectively. Moreover,

$$
\begin{aligned}
g_{ij} &= e^{2\gamma}(\delta_{ij} + \gamma_i \gamma_j), \\
g^{ij} &= e^{-2\gamma}(e^{ij} - \frac{\gamma_i \gamma_j}{\omega^2}) \\
h_{ij} &= \frac{e^\gamma}{\omega}(-\gamma_{ij} + \gamma_i \gamma_j + e_{ij}) \\
h^i_j &= \frac{e^{-\gamma}}{\omega}(e^{ik} - \frac{\gamma_i \gamma_k}{\omega^2})(-\gamma_{kj} + \gamma_k \gamma_j + e_{kj}).
\end{aligned}
\tag{28}
$$

Notice that the Weingarten tensor in (28) is in general not symmetric with respect local lo orthonormal frames $(e_1, \cdots, e_n)$ on $\mathbb{S}^n$, even though it is symmetric with respect to local orthonormal frames on M. We observe that the symmetric matrix $(e^{ij} - \frac{\gamma_i \gamma_j}{\omega^2})$ has an obvious square root S. That is,

$$
S = (S_{ij}) = (e_{ij} - \frac{\gamma_i \gamma_j}{\omega(\omega + 1)}), \quad (e^{ij} - \frac{\gamma_i \gamma_j}{\omega^2}) = S^2.
\tag{29}
$$

S can be used to symmetrize the Weingarten tensor. The eigenvalues of (h^i_j) is the same as eigenvalues of $\frac{e^{-\gamma}}{\omega} B$, with B defined as

$$
\begin{aligned}
B &= : (b_{ij}) = S(-\gamma_{lm} + \gamma_l \gamma_m + e_{lm})S \\
&= (-\gamma_{ij} + \delta_{ij} + \frac{\sum_l (\gamma_i \gamma_{lj} + \gamma_j \gamma_{il}) \gamma_l}{\omega(\omega + 1)} - \frac{\gamma_i \gamma_j \sum_{l,m} \gamma_l \gamma_{lm} \gamma_m}{\omega^2 (1 + \omega)^2}).
\end{aligned}
\tag{30}
$$

Curvature equation (26) can be rewritten as

$$
\frac{e^{(n-k)\gamma}}{\omega^{k-1}} \sigma_k(B) = f.
\tag{31}
$$

As B is a function in $\overline{\nabla}^2 \gamma, \overline{\nabla}\gamma$ only, it is independent of γ. Set

$$
\tilde{F}(\overline{\nabla}^2 \gamma, \overline{\nabla}\gamma) = -\sigma_k(B).
\tag{32}
$$

Denote $\sigma_k^{ij}(B) = \frac{\partial \sigma_k}{\partial b_{ij}}$, we compute

$$
(\tilde{F}^{ij}) = (\frac{\partial \tilde{F}}{\partial \gamma_{ij}}) = S(\sigma_k^{ij}(B))S.
\tag{33}
$$

Since S in (29) is positive definite, we have $(\frac{\partial \tilde{F}}{\partial \gamma_{ij}}) > 0$.

3.1 Uniqueness and C^1-Estimates

Lemma 3.3. *Let* $1 \leq k < n$. *Let* L *denote the linearized operator at a solution* ρ *of (26), if* v *satisfies* $L(v) = 0$ *on* $\mathbb{S}^n$, *then* $v \equiv 0$ *on* $\mathbb{S}^n$. *Moreover, suppose* ρ, $\tilde{\rho}$ *are two solutions of (26) and* $\lambda(\rho_i) \in \Gamma_k$, *for* $i = 1, 2$. *Then* $\rho_1 \equiv \rho_2$.

Proof. (31) can be put in the form of

$$\frac{e^{(n-k)\gamma}}{\omega^{k-1}} \tilde{F}(\overline{\nabla}^2 \gamma, \overline{\nabla}\gamma) = -f. \tag{34}$$

The linearized operator at γ is

$$L(v) = \frac{e^{(n-k)\gamma}}{\omega^{k-1}} \tilde{F}^{ij} v_{ij} + \sum_l b_l v_l - (n-k) f v,$$

for some function $b_l, l = 1, \cdots, n$. The first statement in lemma follows immediately from the maximum principle.

Suppose $\gamma = \log \rho$ and $\tilde{\gamma} = \log \tilde{\rho}$ are two solutions of (31), denote $\tilde{\omega} = \sqrt{1 + |\overline{\nabla}\tilde{\gamma}|^2}$ and $\tilde{B}$ to be the corresponding tensor B in (30) with γ replaced by $\tilde{\gamma}$. For $t \in [0, 1]$, set

$$\gamma^t = t\gamma + (1-t)\tilde{\gamma}, \quad \omega_t = \sqrt{1 + |\overline{\nabla}\gamma^t|^2}, \quad B^t = tB + (1-t)\tilde{B}.$$

Set $v = \gamma - \tilde{\gamma}$, as $B^t \in \Gamma_k$,

$$0 = \frac{e^{(n-k)\gamma}}{\omega^{k-1}} F(B) - \frac{e^{(n-k)\tilde{\gamma}}}{\tilde{\omega}^{k-1}} F(\tilde{B})$$

$$= \int_0^1 \frac{d}{dt} (\frac{e^{(n-k)\gamma^t}}{\omega_t^{k-1}} F(B^t)) dt$$

$$= \int_0^1 (n-k)(\frac{e^{(n-k)\gamma^t}}{\omega_t^{k-1}} F(B^t)) dt + \int_0^1 (\frac{e^{(n-k)\gamma^t}}{\omega_t^{k-1}} F^{ij}(B^t)) dt (b_{ij} - \tilde{b}_{ij}) + mod(\overline{\nabla}v).$$

Write $S = (S_j^i)$, and observe that S only involves $\overline{\nabla}\gamma, \overline{\nabla}^2\gamma$ (and so is $\tilde{S}$), by the Mean Value Theorem,

$$B - \tilde{B} = -S(\overline{\nabla}^2 v)S + mod(\overline{\nabla}v),$$

and

$$0 = (\int_0^1 (n-k)(\frac{e^{(n-k)\gamma^t}}{\omega_t^{k-1}} \tilde{F}(B^t)) dt) v - \int_0^1 (\frac{e^{(n-k)\gamma^t}}{\omega_t^{k-1}} F^{ij}(B^t)) dt) S_i^\alpha S_j^\beta v_{\alpha\beta} + mod(\overline{\nabla}v).$$

Since $(\int_0^1(\frac{e^{(n-k)\gamma^t}}{\omega_t^{k-1}}F^{ij}(B^t))dt)S^{i\alpha}S^{\beta j}) > 0$, $\int_0^1(n-k)(\frac{e^{(n-k)\gamma^t}}{\omega_t^{k-1}}\tilde{F}(B^t))dt > 0$, v satisfies the following elliptic equation,

$$a^{ij}(x)v_{ij}(x) + b^k(x)v_k(x) + c(x)v(x) = 0, \quad \forall x \in \mathbb{S}^n,$$

with $c(x) < 0$ for all $x \in \mathbb{S}^n$. The maximum principle yields $v \equiv 0$. That is $\rho = \tilde{\rho}$.

$\square$

It is useful to write down some differential identities for general C^1 symmetric function F. $F(W)$ is symmetric if it is invariant under orthonormal transformation. With B is defined in (30), set $\tilde{F}(\overline{\nabla}^2\gamma, \overline{\nabla}\gamma) = -F(B)$. Define $F^{ij} = \frac{\partial F}{\partial b_{ij}}$, $\tilde{F}^{ij} = \frac{\partial \tilde{F}}{\partial \gamma_{ij}}$. It follows from (30) that

$$(\tilde{F}^{ij}) = S(F^{ij})S. \tag{35}$$

Lemma 3.4. *For any C^1 symmetric function $F(B)$, set $\phi = \frac{|\overline{\nabla}\gamma|^2}{2}$, then there exist c_m depending on $(\overline{\nabla}^2\gamma, \overline{\nabla}\gamma, F)$, such that*

$$\tilde{F}^{ij}\phi_{ij} = \sum_m c_m\phi_m - \sum_l \gamma_l(F(B))_l + F^{ij}(\delta_{ij}|\overline{\nabla}\gamma|^2 - \gamma_j\gamma_i + \delta_{ij}\gamma_{ii}^2). \tag{36}$$

Proof. By (30),

$$\phi_{ij} = \sum_l (\gamma_l\gamma_{lij} + \gamma_{li}\gamma_{lj})$$

$$= \sum_l (\gamma_l(\gamma_{lij} + \delta_{li}\gamma_j - \gamma_j\delta_{il}) + \gamma_{li}\gamma_{lj})$$

$$= \sum_l (\gamma_l(\gamma_{ijl} + \delta_{ij}\gamma_l - \gamma_j\delta_{il}) + \gamma_{li}\gamma_{lj})$$

$$= \sum_l \gamma_l(-b_{ijl} + (\frac{\gamma_i\phi_j + \gamma_j\phi_i}{\omega(\omega+1)} - \frac{\gamma_i\gamma_j\sum_m\gamma_m\phi_m}{\omega^2(1+\omega)^2})_l)$$

$$+\delta_{ij}|\overline{\nabla}\gamma|^2 - \gamma_j\gamma_i + \delta_{ij}\gamma_{ii}^2$$

$$= \sum_l \gamma_l(-b_{ijl} + (\frac{\gamma_i\phi_{lj} + \gamma_j\phi_{li}}{\omega(\omega+1)} - \frac{\gamma_i\gamma_j\sum_m\gamma_m\phi_{ml}}{\omega^2(1+\omega)^2}))$$

$$+\delta_{ij}|\overline{\nabla}\gamma|^2 - \gamma_j\gamma_i + \delta_{ij}\gamma_{ii}^2 + c_{ij}^m\phi_m,$$

where we used the fact that tensor $A_{ij} := \gamma_{ij} + \gamma e_{ij}$ is Codazzi for any function $\gamma \in C^3(\mathbb{S}^n)$. We rewrite above identity as

$$\phi_{ij} = \sum_l \Big(\frac{\gamma_i \gamma_l \phi_{lj} + \gamma_l \gamma_j \phi_{li}}{\omega(\omega + 1)} - \frac{\gamma_i \gamma_j \sum_{m,l} \gamma_l \gamma_m \phi_{ml}}{\omega^2(1 + \omega)^2} \Big) c_{ij}^m \phi_m$$

$$+ \delta_{ij} |\overline{\nabla}\gamma|^2 - \gamma_j \gamma_i + \delta_{ij} \gamma_{ii}^2 - \sum_l \gamma_l b_{ijl},$$

or equivalently

$$S\overline{\nabla}^2 \phi S - (c_{ij}^m \phi_m) = |\overline{\nabla}\gamma|^2 I - (\gamma_i \gamma_j) + (\overline{\nabla}^2 \gamma)^2 - (\sum_l \gamma_l b_{ijl}).$$

Set $c_m = \sum_{ij} F^{ij} c_{ij}^m$, contracting above identity with F^{ij}, it follows from (35),

$$\tilde{F}^{ij} \phi_{ij} - \sum_m c_m \phi_m = - \sum_l F^{ij}(B) \gamma_l b_{ijl} + F^{ij}(\delta_{ij} |\overline{\nabla}\gamma|^2 - \gamma_j \gamma_i + \delta_{ij} \gamma_{ii}^2)$$

$$= - \sum_l \gamma_l (F(B))_l + F^{ij}(\delta_{ij} |\overline{\nabla}\gamma|^2 - \gamma_j \gamma_i + \delta_{ij} \gamma_{ii}^2).$$

□

Proposition 3.5. *If M satisfies (26), then*

$$\Big(\frac{\min_{\mathbb{S}^n} f}{C_n^k} \Big)^{\frac{1}{n-k}} \le \min_{\mathbb{S}^n} |X| \le \max_{\mathbb{S}^n} |X| \le \Big(\frac{\max_{\mathbb{S}^n} f}{C_n^k} \Big)^{\frac{1}{n-k}}.$$

Moreover, there exits a constant C depending only on n, k, $\min_{\mathbb{S}^n} f$, $|f|_{C^1}$ such that

$$\max_{\mathbb{S}^n} |\overline{\nabla}\rho| \le C.$$

Proof. (γ_{ij}) is semi-negative definite at maximum point of ρ and $\overline{\nabla}\gamma = 0$. By (31),

$$f = \frac{e^{(n-k)\gamma}}{\omega^{k-1}} \sigma_k(B) = e^{(n-k)\gamma} \sigma_k(B) \ge e^{(n-k)\gamma}.$$

This yields an upper bound of γ. A lower bound of γ follows similarly, as (γ_{ij}) is semi-positive definite at any minimum point of ρ.

To obtain an upper bound for $|\overline{\nabla}\rho|$ is now equivalent to obtain an upper bound of $\phi = \frac{|\overline{\nabla}\gamma|^2}{2}$. Suppose $p \in \mathbb{S}^n$ is a maximum point of ϕ. At p,

$$\overline{\nabla}|\overline{\nabla}\gamma|^2 = 0, \quad \overline{\nabla}\omega = 0, \quad B = (-\gamma_{ij} + \delta_{ij}). \tag{37}$$

It follows from (36) with $F(B) = \sigma_k(B)$, at p,

$$0 \geq \sum_{ij} F^{ij} \phi_{ij}$$

$$= -\sum_l \gamma_l (\sigma_k(B))_l + \sum_{ij} \sigma_k^{ij} (\delta_{ij} |\overline{\nabla} \gamma|^2 - \gamma_i \gamma_j + \delta_{ij} \gamma_{ii}^2)$$

$$\geq -\sum_l \gamma_l (e^{-(n-k)\gamma} \omega^{k-1} f)_l$$

$$= ((n-k)|\overline{\nabla} \gamma|^2 f - \overline{\nabla} \gamma \cdot \overline{\nabla} f) e^{(k-n)\gamma} \omega^{k-1}$$

$$\geq c(|\overline{\nabla} \gamma|^2 - C|\overline{\nabla} \gamma|) e^{(k-n)\gamma} \omega^{k-1}, \tag{38}$$

where $c \geq \delta, C \leq \frac{1}{\delta}$ are two positive constants with δ depending only on $n, k, \inf f, |\overline{\nabla} f|$. The gradient estimate follows from (38). $\qquad\square$

3.2 C^2-Estimates and the Existence

We precede to prove C^2 a priori estimates, this is equivalent to obtain curvature estimate for M due to C^1 estimates we have already obtained. For this purpose, it is convenient to work directly on induced metric g on $M \subset \mathbb{R}^{n+1}$. For $X \in M$, choose local orthonormal frame $\{e_1, \cdots, e_n\}$ on M, and $\nu = e_{n+1}$ is the unit outer normal of the hypersurface, such that $\{e_1, \cdots, e_{n+1}\}$ of $\mathbb{R}^{n+1}$ is a local orthonormal frame in $\mathbb{R}^{n+1}$. We use lower indices to denote covariant derivatives with respect to the induced metric.

The second fundamental form is the symmetric $(2, 0)$-tensor given by the matrix $\{h_{ij}\}$, and we denote the Weingarten tensor $\{h_i^j\} = \{g^{jl} h_{li}\}$,

$$h_{ij} = \langle \partial_i X, \partial_j \nu \rangle. \tag{39}$$

We have the following identities,

$$\begin{aligned} X_{ij} &= -h_{ij}\nu \quad \text{(Gauss formula)} \\ (\nu)_i &= h_i^j X_j \quad \text{(Weigarten equation)} \\ h_{ijk} &= h_{ikj} \quad \text{(Codazzi formula)} \\ R_{ijkl} &= h_{ik} h_{jl} - h_{il} h_{jk} \quad \text{(Gauss equation)}, \end{aligned} \tag{40}$$

where R_{ijkl} is the $(4, 0)$-Riemannian curvature tensor. We also have

$$\begin{aligned} h_{ijkl} &= h_{ijlk} + h_{mj} R_{imlk} + h_{im} R_{jmlk} \\ &= h_{klij} + (h_{mj} h_{il} - h_{ml} h_{ij}) h_{mk} + (h_{mj} h_{kl} - h_{ml} h_{kj}) h_{mi}. \end{aligned} \tag{41}$$

Since $\{e_1, \cdots, e_n\}$ is an orthonormal frame on M, $g_{ij} = \delta_{ij}$, $h_{ij} = h_j^i$. The principal curvatures $(\kappa_1, \cdots, \kappa_n)$ are the eigenvalues of the second fundamental form with respect to the metric which satisfy

$$\det(h_{ij} - \kappa g_{ij}) = 0.$$

The curvature equation (26) on $\mathbb{S}^n$ can also be equivalently expressed as a curvature equation on M,

$$\sigma_k(\kappa_1, \cdots, \kappa_n)(X) = \frac{u(X)}{|X|^{n+1}} f\left(\frac{X}{|X|}\right), \quad \forall X \in M. \tag{42}$$

Proposition 3.6. *For $1 < k < n$, let $F \equiv \sigma_k = \Phi u$ and denote $H \equiv \sigma_1$, then at a maximum point of $\frac{H}{u}$,*

$$\begin{aligned} F^{ij}\left(\frac{H}{u}\right)_{ij} &= \frac{1}{u}[\Phi_{ss}u + 2\Phi_s u_s] - \left(\frac{H}{u}\right)\Phi_l\langle X, X_l\rangle - (k-1)\left(\frac{H}{u}\right)\Phi \\ &\quad + (k-1)\phi|A|^2 - \frac{1}{u}F^{ij;ml}h_{ij;s}h_{ml;s}, \end{aligned} \tag{43}$$

where A denotes the second fundamental form.

Proof. By definition, $u = \langle X, \nu\rangle$. Compute the first and second order covariant derivatives, we have

$$\begin{aligned} u_s &= h_{sl}\langle X, X_l\rangle \\ u_{ij} &= h_{ij;l}\langle X, X_l\rangle + h_{ij} - (h^2)_{ij}u \end{aligned} \tag{44}$$

Also since (h_{ij}) is Codazzi, by Ricci identity and Gauss equation,

$$\begin{aligned} h_{ij;kl} &= h_{kl;ij} + (h_{lk}h_{im} - h_{lm}h_{ik})h_{mj} + (h_{lj}h_{im} - h_{lm}h_{ij})h_{mk} \\ F^{ij}h_{ij;st} &= F_{st} - F^{ij;ml}h_{ml;s}h_{ij;t}. \end{aligned} \tag{45}$$

At any maximum point $P \in M^n$ of $\frac{H}{u}$, $\left(\frac{H}{u}\right)_i(P) = 0$. At P,

$$\begin{aligned} F^{ij}\left(\frac{H}{u}\right)_{ij} &= F^{ij}\left[\frac{H_{ij}}{u} - \frac{u_j}{u}\left(\frac{H}{u}\right)_i - \frac{u_i}{u}\left(\frac{H}{u}\right)_j - \left(\frac{H}{u}\right)\frac{u_{ij}}{u}\right] \\ &= \frac{1}{u}F^{ij}H_{ij} - \frac{1}{u}\left(\frac{H}{u}\right)F^{ij}u_{ij}. \end{aligned} \tag{46}$$

Apply formulas (44) and (45),

$$
\begin{aligned}
\tfrac{1}{u} F^{ij} H_{ij} &= \tfrac{1}{u} F^{ij} h_{ss;ij} \\
&= \tfrac{1}{u} F^{ij} \big[h_{ij;ss} + (h_{ij}h_{sm} - h_{jm}h_{si})h_{ms} + (h_{js}h_{sm} - h_{jm}h_{ss})h_{mi} \big] \\
&= \tfrac{1}{u} F^{ij} h_{ij;ss} + k\Phi |A|^2 - \tfrac{1}{u} F^{ij} (h^2)_{ij} H \\
&= \tfrac{1}{u} F_{ss} - \tfrac{1}{u} F^{ij;ml} h_{ij;s} h_{ml;s} + k\Phi |A|^2 - \big(\tfrac{H}{u} \big) F^{ij} (h^2)_{ij} \\
&= \tfrac{1}{u} [\Phi_{ss} u + 2\Phi_s u_s + \Phi u_{ss}] - \tfrac{1}{u} F^{ij;ml} h_{ij;s} h_{ml;s} + k\Phi |A|^2 \\
&\quad - \big(\tfrac{H}{u} \big) F^{ij} (h^2)_{ij} \\
&= \tfrac{1}{u} [\Phi_{ss} u + 2\Phi_s u_s] + \tfrac{\Phi}{u} \big[H_l \langle X, X_l \rangle + H - |A|^2 u \big] \\
&\quad - \tfrac{1}{u} F^{ij;ml} h_{ij;s} h_{ml;s} + k\Phi |A|^2 - \big(\tfrac{H}{u} \big) F^{ij} (h^2)_{ij} \\
&= \tfrac{1}{u} [\Phi_{ss} u + 2\Phi_s u_s] + \tfrac{\Phi}{u} H_l \langle X, X_l \rangle + \big(\tfrac{H}{u} \big) \Phi \\
&\quad - \tfrac{1}{u} F^{ij;ml} h_{ij;s} h_{ml;s} + (k-1)\phi |A|^2 - \big(\tfrac{H}{u} \big) F^{ij} (h^2)_{ij}.
\end{aligned}
\tag{47}
$$

We also compute

$$
\begin{aligned}
-\tfrac{1}{u} \big(\tfrac{H}{u} \big) F^{ij} u_{ij} &= -\tfrac{1}{u} \big(\tfrac{H}{u} \big) F^{ij} \Big[h_{ij;l} \langle X, X_l \rangle + h_{ij} - (h^2)_{ij} u \Big] \\
&= -\tfrac{1}{u} \big(\tfrac{H}{u} \big) F_l \langle X, X_l \rangle - k\phi \big(\tfrac{H}{u} \big) + \big(\tfrac{H}{u} \big) F^{ij} (h^2)_{ij} \\
&= -\tfrac{\Phi}{u} \big(\tfrac{H}{u} \big) u_l \langle X, X_l \rangle - \big(\tfrac{H}{u} \big) \Phi_l \langle X, X_l \rangle - k\Phi \big(\tfrac{H}{u} \big) + \big(\tfrac{H}{u} \big) F^{ij} (h^2)_{ij},
\end{aligned}
\tag{48}
$$

where $(h^2)_{ij} = h_{ik}h_{kj}$.

Adding up (47) and (48), and using the critical point condition, we obtain

$$
\begin{aligned}
F^{ij} \big(\tfrac{H}{u} \big)_{ij} &= \tfrac{1}{u} [\Phi_{ss} u + 2\Phi_s u_s] + \phi \big(\tfrac{H}{u} \big)_l \langle X, X_l \rangle - \big(\tfrac{H}{u} \big) \Phi_l \langle X, X_l \rangle \\
&\quad - (k-1)\big(\tfrac{H}{u} \big) \Phi - \tfrac{1}{u} F^{ij;ml} h_{ij;s} h_{ml;s} + (k-1)\Phi |A|^2 \\
&= \tfrac{1}{u} [\Phi_{ss} u + 2\Phi_s u_s] - \big(\tfrac{H}{u} \big) \Phi_l \langle X, X_l \rangle - (k-1)\big(\tfrac{H}{u} \big) \Phi \\
&\quad - \tfrac{1}{u} F^{ij;ml} h_{ij;s} h_{ml;s} + (k-1)\Phi |A|^2,
\end{aligned}
\tag{49}
$$

(43) is verified.   □

C^2 estimates can be established with the help of Proposition 3.6 and Corollary 2.9.

Lemma 3.7. *If M satisfies (42) for some $1 \le k \le n$, then there exists a constant C depending only on n, k, $\min_{S^n} f$, $|f|_{C^1}$, and $|f|_{C^2}$, such that*

$$
\max_M \sigma_1 \le C, \qquad |\nabla^2 \rho| \le C.
\tag{50}
$$

Proof. We have already obtained the C^0 and C^1 estimates for ρ. For the case of $k = 1$, (42) is a mean curvature type equation which is of divergent form of quasilinear PDE. C^2 estimates follows from the classical quasilinear elliptic PDE theory. We work on $2 \leq k \leq n - 1$ cases. When $k > 1$, the estimation of the curvature bound is equivalent to the estimation of mean curvature H (which yields C^2 bound on ρ). To see this, suppose mean curvature $H \leq C$ is bounded from above. Since $\kappa \in \Gamma_k \subset \Gamma_2$, $(\kappa|i) \in \Gamma_1$. Hence, for each i,

$$C \geq H = \sigma_1(\kappa) = \kappa_i + \sigma_1(\kappa|i) \geq \kappa_i.$$

This give an upper bound of curvature. A lower bound follows from the fact $\sigma_1(\kappa) > 0$ and $\kappa_i \leq C$ for each i.

As u is bounded from below and above, we only need to get an upper bound of $\frac{H}{u}$. Suppose $P \in M$ where $\frac{H}{u}$ achieves its maximum, it follows from (43)

$$
\begin{aligned}
0 \geq F^{ij}\left(\tfrac{H}{u}\right)_{ij} \\
= \tfrac{1}{u}[\Phi_{ss}u + 2\phi_s u_s] - \left(\tfrac{H}{u}\right)\Phi_l\langle X, X_l\rangle - (k-1)\left(\tfrac{H}{u}\right)\Phi \\
- \tfrac{1}{u}F^{ij;ml}h_{ij;s}h_{ml;s} + (k-1)\Phi|A|^2.
\end{aligned}
\tag{51}
$$

Recall $\Phi(X) = |X|^{-(n+1)} f(\frac{X}{|X|})$ and with C^0, C^1 estimates of $\rho = |X|$, we have the following estimates.

$$|\Phi_i|(P) \leq C(n, k, \min_{S^n} f, |f|_{C^1})$$
$$|\Phi_{ii}|(P) \leq C(n, k, \min_{S^n} f, |f|_{C^1}, |f|_{C^2})\big(1 + |A|(P)\big)$$

On the other hand, $|u_i| = |h^i_j \rho \rho_j| \leq c_3|A|$. By (42),

$$\frac{\sigma_1}{u} = \frac{\sigma_1 \phi}{\sigma_k}.$$

At a maximum point P of the test function $\frac{\sigma_1}{u}$, one has

$$\frac{(\sigma_1)_s}{\sigma_1} = \frac{(\sigma_k)_s}{\sigma_k} - \frac{\phi_s}{\phi}.$$

In Corollary 2.9, set $r = \frac{\phi_s}{\phi}(P)$, then

$$
\begin{aligned}
F^{ij;ml}h_{ij;s}h_{ml;s} &\leq 2r(u\phi)_s - \tfrac{k}{k-1}r^2 u\phi \\
&\leq C_1(n, k, \min_{S^n} f, |f|_{C^1})|A| + C_2(n, k, \min_{S^n} f, |f|_{C^1}).
\end{aligned}
$$

With the above estimates, (51) can be simplified as

$$|A|^2(P) + c_4|A|(P) + c_5 \leq 0,
\tag{52}$$

where c_4 and c_5 are constants depending only on n, k, $\min_{S^n} \phi$, $|f|_{C^1}$, and $|f|_{C^2}$. Hence at P, $|A|(P) \leq C$. In turn

$$\sigma_1(X) \leq u(X)\frac{\sigma_1(P)}{u(P)} \leq C, \quad \text{for any } X \in M.$$

This implies (50). $\square$

We prove Theorem 3.2 using the method of continuity.

Proof. For any positive function $f \in C^2(\mathbb{S}^n)$, for $0 \leq t \leq 1$ and $1 \leq k < n-1$, set

$$f_t(x) = [1 - t + tf^{-\frac{1}{k}}(x)]^{-k}.$$

Consider the following family of equations for $0 \leq t \leq 1$:

$$\sigma_k^{\frac{1}{k}}(\kappa_1, \cdots, \kappa_n)(x) = (f_t(x)\rho^{1-n}(\rho^2 + |\nabla\rho|^2)^{-1/2})^{\frac{1}{k}}, \quad \text{on } \mathbb{S}^n, \qquad (53)$$

where $n \geq 2$. We want to find admissible solutions in the class of star-shaped hypersurfaces. Set

$$I = \{t \in [0, 1] : \text{such that (53) is solvable.}\}$$

I is nonempty because $\rho = [C_n^k]^{-\frac{1}{n-2}}$ is a solution for $t = 0$. By Lemmas 3.5, 3.7, 2.6 and 2.7, equation (53) is unform elliptic and concave, apply the Evans-Krylov theorem and the Schauder theorem, we have

$$\|\rho\|_{C^{3,\alpha}(\mathbb{S}^n)} \leq C,$$

where C depends only on only on n, k, $\min_{S^n} f$, $\max_{S^n} f$, $|f|_{C^1}$, $|f|_{C^2}$ and α. The a priori estimates guarantee that I is closed. The openness comes from Lemma 3.3 and the inverse function theorem. This proves the existence part of the theorem. The uniqueness part of the theorem follows from Lemma 3.3. $\square$

4 Isoperimetric Inequality for Quermassintegrals on Starshaped Domains

In this section, we use a geometric flow to establish isoperimetric inequalities for quermassintegrals of k-convex starshaped domains in $\mathbb{R}^{n+1}$.

Theorem 4.1. *Suppose $1 \leq n-1$, and suppose Ω is a k-convex starshaped domain in $\mathbb{R}^{n+1}$, then the following inequality holds,*

$$(V_{(n+1)-k}(\Omega))^{\frac{1}{n+1-k}} \leq \mathbf{C}_{n,k}(V_{n-k}(\Omega))^{\frac{1}{n-k}}, \qquad (54)$$

where

$$\mathbf{C}_{n,k} = \frac{(V_{(n+1)-k}(B))^{\frac{1}{n+1-k}}}{(V_{n-k}(B))^{\frac{1}{n-k}}},$$

B is the standard ball in $\mathbb{R}^{n+1}$. The equality holds if and only if Ω is a ball.

We consider the following normalized evolution equation on hypersurface M^n in $\mathbb{R}^{n+1}$.

$$\partial_t X = (\frac{1}{F(\kappa)} - ru)v, \tag{55}$$

where $F(\cdot, t)$ and $r(t)$ are to be determined, $u = <X, v>$ is the supporting function of the hypersurface.

We derive the evolution equations of various geometric quantities for the following general flow.

$$\partial_t X = fv. \tag{56}$$

Proposition 4.2. *Under flow (56), the following evolution equations hold.*

$$\begin{aligned}
\partial_t g_{ij} &= 2f h_{ij} \\
\partial_t v &= -\nabla f \\
\partial_t h_{ij} &= -\nabla_i \nabla_j f + f(h^2)_{ij} \\
\partial_t h^i_j &= -\nabla^i \nabla_j f - f(h^2)^i_j \\
\partial_t \sigma_k &= -\sum_{ij} \sigma_k^{ij}(g^{-1}h) f_{ij} - f\left(\sigma_1(g^{-1}h)\sigma_k(g^{-1}h) - (k+1)\sigma_{k+1}(g^{-1}h)\right)
\end{aligned} \tag{57}$$

Proof. Pick any local coordinate chart $(x_1, \cdots, x_n)$ of M, denote $X_i = \frac{\partial X}{\partial x_i}$, $i = 1, \cdots, n$, as $\langle X_i, v \rangle = 0$, $\forall i$, by Weingarten equation (40),

$$\begin{aligned}
(g_{ij})_t &= \langle X_i, X_j \rangle_t \\
&= \langle X_{i,t}, X_j \rangle + \langle X_i, X_{j,t} \rangle \\
&= \langle X_{t,i}, X_j \rangle + \langle X_i, X_{t,j} \rangle \\
&= \langle (fv)_i, X_j \rangle + \langle X_i, (fv)_j \rangle \\
&= f \langle (v)_i, X_j \rangle + f \langle X_i, (v)_j \rangle \\
&= f \langle \sum_l h^l_i X_l, X_j \rangle + f \langle X_i, \sum_l h^l_j X_l \rangle \\
&= f \sum_l h^l_i g_{lj} + f \sum_l h^l_j g_{li} \\
&= 2f h_{ij}
\end{aligned}$$

Since v is a unit vector field, v_t has only tangential component. We only need to compute $\langle v_t, X_i \rangle$. As $\langle v, X_i \rangle \equiv 0$,

$$\langle v_t, X_i \rangle = -\langle v, X_{i,t} \rangle = -\langle v, (fv)_i \rangle = -\langle v, (f)_i v \rangle = -f_i.$$

This verifies the second identity in the proposition.

For the third identity, again using the fact v is a unit vector field, by the second identity we just proved and the Gauss formula in (40),

$$\begin{aligned}
h_{ij,t} &= -\langle X_{ij}, v \rangle_t \\
&= -\langle X_{ij,t}, v \rangle - \langle X_{ij}, v_t \rangle \\
&= -\langle (fv)_{ij}, v \rangle + \langle h_{ij}v, \nabla f \rangle \\
&= -f_{ij} - f \langle v_{ij}, v \rangle \\
&= -f_{ij} - f \langle (h_i^l X_l)_j, v \rangle \\
&= -f_{ij} - f \langle (h_i^l)_j X_l v \rangle - f \langle h_i^l X_{lj}, v \rangle \\
&= -f_{ij} + f \langle h_i^l h_{lj} v, v \rangle \\
&= -f_{ij} + f h_i^l h_{lj}.
\end{aligned}$$

The fourth identity follows from the first and third, and the fact $g_t^{ij} = -g^{il} g^{mj} g_{lm,t}$. The final identity in the proposition follows from the fourth identity and Proposition 2.2.   □

Corollary 4.3. *Under flow (55), where F is homogeneous of degree 1, then we have the following evolution equations.*

$$\begin{aligned}
\partial_t g_{ij} &= 2(\frac{1}{F} - ru)h_{ij} \\
\partial_t v &= -\nabla(\frac{1}{F} - ru) \\
\partial_t h_{ij} &= -\nabla_i \nabla_j (\frac{1}{F} - ru) + (\frac{1}{F} - ru)(h^2)_{ij} \\
\partial_t h_j^i &= -\nabla^i \nabla_j (\frac{1}{F} - ru) - (\frac{1}{F} - ru)(h^2)_j^i \\
\partial_t \sigma_k &= -\sum_{ij} \sigma^{ij} \nabla_j \nabla_i (\frac{1}{F} - ru) - (\frac{1}{F} - ru)\sigma_{k-1;i}\lambda_i^2 \\
\partial_t F &= -\dot{F}^{ij} \nabla^i \nabla_j (\frac{1}{F} - ru) - (\frac{1}{F} - ru)\dot{F}^{ij}(h^2)_j^i
\end{aligned} \qquad (58)$$

Furthermore, the following heat type evolution equation for Weingarten map h_j^i is valid.

Proposition 4.4.

$$\partial_t h^i_j = \frac{1}{F^2} \dot{F}^{kl} \nabla^k \nabla_l h^i_j + \frac{1}{F^2} \dot{F}(h^2) h^i_j + \frac{1}{F^2} \ddot{F}(\nabla h, \nabla h)$$
$$- \frac{2}{F^3} \nabla^i F \nabla_j F - \frac{2}{F}(h^2)^i_j + r \nabla^i h^l_j < \nabla_l X, X > + r h^i_j. \tag{59}$$

Proof. It follows from previous corollary, (41) and (44).   □

4.1   Monotonicity Properties

We want to choose F and r in flow (55) such that the corresponding global geometric quantities are monotone along the flow. The Minkowski identity (4) plays key role here.

From identities in Corollary 4.3, for $1 \le l \le n - 1$,

$$\partial_t \int_M \sigma_l d\mu_g = \int_M \partial_t \sigma_l + \sigma_l \frac{1}{2} g^{ij} \partial_t g_{ij} d\mu_g$$
$$= - \int_M (\frac{1}{F} - ru)(\sum_i \sigma_{l-1;i} \lambda_i^2 - \sigma_l \sigma_1) d\mu_g$$
$$= (l + 1) \int_M (\frac{1}{F} - ru)\sigma_{l+1} d\mu_g \tag{60}$$
$$= (l + 1)\left[\int_M \frac{1}{F}\sigma_{l+1} d\mu_g - r \int_M u\sigma_{l+1} d\mu_g\right]$$
$$= (l + 1)\left[\int_M \frac{1}{F}\sigma_{l+1} d\mu_g - rC_{n,l} \int_M \sigma_l d\mu_g\right],$$

where $C_{n,l} = \frac{\sigma_{l+1}(I)}{\sigma_l(I)}$ is the constant in the Minkowski equality.

For the special case $l = n$ and for any f, by Proposition 4.2, along flow (56),

$$\partial_t \int_M \sigma_n d\mu_g = \int_M \partial_t \sigma_n + \sigma_n \frac{1}{2} g^{ij} \partial_t g_{ij} d\mu_g$$
$$= - \int_M f(\sum_i \sigma_{n-1;i} \lambda_i^2 - \sigma_n \sigma_1) d\mu_g \tag{61}$$
$$= (l + 1) \int_M f(\sigma_n \sigma_1 - \sigma_n \sigma_1) d\mu_g$$
$$= 0$$

That is, $V_0(\Omega)$ is a topological invariant. This gives topological obstruction for the problem of prescribing curvature measure $\mathcal{C}_0$.

From (60), if one wants to fix $\int_M \sigma_k d\mu_g$, one may choose $F = \frac{\sigma_k}{\sigma_{k-1}}$ in (55) and define r as

$$r(t) = \frac{\int_{M_t} \frac{\sigma_{k+1}\sigma_{k-1}}{\sigma_k} d\mu_g}{C_{n,k} \int_M \sigma_k d\mu_g}. \tag{62}$$

To be precise, we consider the normalized flow

$$\partial_t X = \left(\frac{\sigma_{k-1}}{\sigma_k} - ru \right) v, \tag{63}$$

The first step is to get an estimate on $r(t)$.

Lemma 4.5. *$r(t)$ is invariant under rescaling, and*

$$r(t) \leq (\frac{\sigma_{k-1}}{\sigma_k})(I) = C_{n,k-1}, \tag{64}$$

equality holds if and only if M_t is the standard sphere.

Proof. The inequality follows directly from the Newton-MacLaurin inequality. If the equality holds, this means the Newton-MacLaurin inequality holds at every point of M_t. So M_t is umbilical at every point, it is a sphere. □

The following monotonicity property is crucial.

Proposition 4.6. *For any k-convex domain Ω, under flow equation (63), we have*

1. $\int_M \sigma_k d\mu_g$ *is a constant;*

2. $\int_M \sigma_{k-1} d\mu_g$ *is monotonically non-decreasing.*

Proof. By the choice of r and (60),

$$\partial_t \int_M \sigma_k d\mu_g = 0. \tag{65}$$

This proves the first part of the statement.
From (60),

$$\begin{aligned}
\partial_t \int_M \sigma_{k-1} d\mu_g &= k \left[\int_M \frac{1}{F} \sigma_k d\mu_g - r C_{n,k-1} \int_M \sigma_{k-1} d\mu_g \right] \\
&= k \int_M \left[\frac{1}{F} \frac{\sigma_k}{\sigma_{k-1}} - r C_{n,k-1} \right] \sigma_{k-1} d\mu_g \\
&= k \int_M \left[1 - \frac{\int_M \frac{\sigma_{k+1}\sigma_{k-1}}{\sigma_k} d\mu_g}{C_{n,k} \int_M \sigma_k d\mu_g} C_{n,k-1} \right] \sigma_{k-1} d\mu_g \\
&\geq k \int_M \left[1 - \frac{\sigma_{k+1}(I)\sigma_{k-1}(I)}{\sigma_k^2(I)} \frac{C_{n,k-1}}{C_{n,k}} \right] \sigma_{k-1} d\mu_g = 0,
\end{aligned} \tag{66}$$

where we used the Newton-MacLaurine inequality in the last step. □

We want to establish the following longtime existence and convergence of flow (63).

Theorem 4.7. *If Ω_0 is k-convex starshaped domain with smooth boundary M_0, flow (63) exists all time $t > 0$, it converges to a standard sphere centered at the origin.*

By a proper rescaling, we will assume $V_k(\Omega_0) = V_k(B)$ where B is the standard ball in $\mathbb{R}^{n+1}$.

The rest of the section is devoted to the proof of Theorem 4.7.

4.2 The Harnack Estimate

If M^n is starshaped, it can be parametrized as $X = \rho(x)x$, where $x \in S^n$. All the geometric information of the hypersurface except the parametrization are encoded in the function $\rho(x)$.

Write $\rho = |X(t)| = \rho(x(t), t)$, where X evolves according to

$$X_t = f\nu.$$

ρ satisfies

$$\frac{d\rho}{dt} = \rho_t + \rho_x \cdot x_t.$$

By (25),

$$\nu = \frac{\rho x - \overline{\nabla}\rho}{\sqrt{\rho^2 + |\overline{\nabla}\rho|^2}}.$$

We have,

$$f\frac{\rho x - \overline{\nabla}\rho}{\sqrt{\rho^2 + |\overline{\nabla}\rho|^2}} = \nu f = X_t = (\rho x)_t = (\rho_t + \rho_x \cdot x_t)x + \rho x_t. \qquad (67)$$

Note that $x_t \perp x$, equalize the tangential components of $\mathbb{S}^n$ in (67),

$$x_t = -\frac{f\overline{\nabla}\rho}{\rho\sqrt{\rho^2 + |\overline{\nabla}\rho|^2}}.$$

Therefore,

$$\rho_x \cdot x_t = \overline{\nabla}\rho \cdot x_t = -\frac{f|\overline{\nabla}\rho|^2}{\rho\sqrt{\rho^2 + |\overline{\nabla}\rho|^2}}.$$

Put the above identity to (67), equalize the normal component of $\mathbb{S}^n$ in (67),

$$\rho_t = -\rho_x \cdot x_t + \frac{f\rho}{\sqrt{\rho^2 + |\overline{\nabla}\rho|^2}} = \frac{f\sqrt{\rho^2 + |\overline{\nabla}\rho|^2}}{\rho}.$$

In particular, if X satisfies (55), ρ satisfies

$$\partial_t \rho = \frac{\sqrt{\rho^2 + |\overline{\nabla}\rho|^2}}{\rho} \frac{1}{F} - r\rho. \tag{68}$$

Equation (68) is equivalent to (55) up to diffeomorphism, if we can prove that the starshapedness is preserved along the flow.

For the gradient estimate, we prefer to work on (68). As in the previous section dealing to the problem of prescribing curvature measure, let $\gamma \equiv \ln \rho$, and we choose a local orthonormal frame $\{e_i\}_{i=1}^n$ on S^n.

By the homogeneity of F,

$$\partial_t \gamma = \frac{\omega^2}{F(B)} - r, \tag{69}$$

where

$$\omega = \sqrt{1 + |\overline{\nabla}\gamma|^2}, \quad B = (-\gamma_{ij} + \delta_{ij} + \frac{\sum_l (\gamma_i \gamma_{lj} + \gamma_j \gamma_{li})\gamma_l}{\omega(\omega+1)} - \frac{\gamma_i \gamma_j \sum_{l,m} \gamma_l \gamma_{lm}\gamma_m}{\omega^2(1+\omega)^2}),$$

as defined in (30).

Proposition 4.8. *Let* $\phi = \frac{|\overline{\nabla}\gamma|^2}{2}$, *assume (69) preserves* $\kappa(t) \in \Gamma_k$,

$$\partial_t \phi = \mathcal{L}_{ij} \overline{\nabla}_i \overline{\nabla}_j \phi + W_k \cdot \overline{\nabla}_k \phi - \frac{\omega^2}{F^2(B)} \sum_{ij} \frac{\partial F}{\partial b_{ij}} (\delta_{ij}|\overline{\nabla}\gamma|^2 - \gamma_j \gamma_i + \delta_{ij}\gamma_{ii}^2). \tag{70}$$

where W_k *is a one-parameter family of vector fields depending on time, and* $\mathcal{L}_{ij}$ *is an elliptic operator defined as follows,*

$$\mathcal{L}_{ij} \equiv \frac{\omega^2}{F^2(B)} \tilde{F}^{ij}, \tag{71}$$

where $\tilde{F}^{ij}$ *defined as in (35). In consequence,* $\overline{\nabla}\gamma$ *is bounded from above independent of time* t.

Proof. $\kappa \in \Gamma_k$ is equivalent to $B \in \Gamma_k$, hence $F(B) > 0$. Rewrite the last equation in (69) as

$$F(B) = \frac{\omega^2}{\gamma_t + r}.$$

Proposition follows from Lemma 3.4 with a straightforward computation using identity (36).   □

The following Harnack type gradient estimate is a directly consequence.

Corollary 4.9. *Let ρ be a positive solution to (68) on $S^n \times [0, T)$. Then there exists a constant C which depends on $\rho(\cdot, 0)$ but independent of t, such that at each time $t \in [0, T)$,*

$$\max_{S^n} \rho(\cdot, t) \leq C \cdot \min_{S^n} \rho(\cdot, t) \tag{72}$$

Proof. We prove the corollary for each fixed time $t_0 \in [0, T)$. Assume $\rho(\cdot, t_0)$ achieves maximum at x_+ and minimum at x_-, and let $\Gamma : [s_1, s_2] \longrightarrow M^n$ be a path joining x_- and x_+. We have

$$
\begin{aligned}
\log \frac{\rho(x_+, t_0)}{\rho(x_-, t_0)} &= \int_{s_1}^{s_2} \frac{d}{ds}[\log \rho(\Gamma(s), t_0)]ds \\
&= \int_{s_1}^{s_2} \frac{\overline{\nabla}\rho}{\rho} \cdot \frac{d\Gamma}{ds}ds \\
&\leq \int_{s_1}^{s_2} \overline{\nabla}\gamma \cdot \left|\frac{d\Gamma}{ds}\right|ds \\
&\leq \tilde{C} \int_{s_1}^{s_2} \left|\frac{d\Gamma}{ds}\right|ds.
\end{aligned}
\tag{73}
$$

By taking Γ to be the shortest geodesic with constant speed 1 which joins x_- and x_+, we obtain $\int_{s_1}^{s_2} \left|\frac{d\Gamma}{ds}\right|ds = d(x_-, x_+) \leq \pi$.   □

Lemma 4.10. *Suppose that $\rho > 0$ satisfies (68), then at any time $t_0 \geq 0$, if $x_0 \in \mathbb{S}^n$ is a minimum point of $\rho(x, t_0)$, then $\rho(x_0, t_0)_t \geq 0$, strict inequality holds unless $M(t_0)$ is a round unit sphere at the origin.*

Proof. The minimum point of $\rho(x, t_0)$ is the same as minimum point of $\gamma(x, t_0)$. By (69),

$$\gamma_t(x_0, t_0) = \frac{\omega^2(x_0, t_0)}{F(B(x_0, t_0))} - r(t_0).$$

As x_0 is a minimum point, $\overline{\nabla}\gamma(x_0, t_0) = 0$, so at (x_0, t_0), $\omega = 1$ and

$$B = (-\overline{\nabla}^2\gamma + I) \leq I. \tag{74}$$

Hence,

$$F(B(x_0, t_0)) \leq F(I). \tag{75}$$

That is,

$$\frac{\omega^2(x_0, t_0)}{F(B(x_0, t_0))} \geq \frac{1}{F(I)}. \tag{76}$$

By Lemma 4.5, $r(t_0) < \dfrac{1}{F(I)}$ unless $M(t_0)$ is a round sphere (by normalization, it is a sphere of radius 1). We have

$$\gamma_t(x_0, t_0) = \frac{\omega^2(x_0, t_0)}{F(B(x_0, t_0))} - r(t_0) \geq \frac{1}{F(I)} - r(t_0) > 0,$$

unless $M(t_0)$ is a round sphere of radius 1. We claim if $\gamma_t(x_0, t_0) = 0$, this round sphere must centered at the origin. Suppose its center z is not the origin, we may assume $z = (0, \cdots, 0, s)$ for some $-1 < s < 0$. Now

$$\gamma(x, t_0) = \frac{1}{2} \log(1 + s^2 + 2sx_{n+1}).$$

The minimum point is $x_0 = (0, \cdots, 0, 1)$, it is easy to compute that

$$-\overline{\nabla}^2 \gamma(x_0, t_0) = \frac{s}{(1+s)^2} I.$$

The strictly inequalities will occur in (74)–(76). Thus,

$$\gamma_t(x_0, t_0) = \frac{\omega^2(x_0, t_0)}{F(B(x_0, t_0))} - r(t_0) > \frac{1}{F(I)} - r(t_0) = 0.$$

contradiction.   □

The following C^0 estimate is a direct consequence of Corollary 4.9 and Lemma 4.10.

Corollary 4.11. *Let ρ be a positive solution to (68) on $S^n \times [0, T)$. Then there exists a uniform positive constant C which does not depend on time t, such that for $\forall t \in [0, T)$,*

$$0 < \tfrac{1}{C} \leq \rho(x, t) \leq C, \tag{77}$$

for any point $(x, t) \in S^n \times [0, T)$. Moreover, $u(x, t) \geq c > 0$ for some constant c independent of t.

Proof. By Lemma 4.10, $\rho(x_0, t_0)_t > 0$ at any minimum point x_0 of $\rho(x, t_0)$, unless $M(t_0)$ is a round unit sphere centered at 0. That is, $\min_{x \in S^n} \rho(x, t)$ is strictly increasing at t_0 unless $M(t_0)$ is a round sphere centered at 0. In any case,

$$\min_{x \in S^n} \rho(x, t) \geq \min_{x \in S^n} \rho(x, 0). \tag{78}$$

An upper bound of ρ follows from the Harnack inequality (72).

The last statement in lemma follow from the identity

$$u = \frac{\rho^2}{\sqrt{\rho^2 + |\overline{\nabla}\rho|^2}}.$$

□

Since u is bound from below by a positive constant independent of t, flow (68) preserves the starshapedness. We want to show that Γ_k is also preserved along the flow. From the property of Γ_k, we only need to show $\sigma_k > 0$ is preserved. This is equivalent to show $F > 0$ is preserved.

Proposition 4.12. *There is $C > 0$, such that $\frac{1}{F} \le C$.*

Proof. We consider function $G = \gamma_t + r$. We may rewrite (69) as

$$G = \frac{\omega^2}{F(B)} =: \overline{F}(\overline{\nabla}\gamma, \overline{\nabla}^2\gamma), \qquad (79)$$

where $(\overline{F}^{ij}) = (\frac{\partial \overline{F}}{\partial \gamma_{ij}}) > 0$. Differentiate (79) in t variable, and notice that r is independent of x,

$$G_t = \sum_{ij} \overline{F}^{ij}(\gamma_t)_{ij} + \sum_l \frac{\partial \overline{F}}{\partial \gamma_l}(\gamma_t)_l$$

$$= \sum_{ij} \overline{F}^{ij}G_{ij} + \sum_l \frac{\partial \overline{F}}{\partial \gamma_l}G_l.$$

G is bounded from above by the maximum principle. Since r is bounded, $\frac{1}{F(B)}$ is bounded. The boundedness of $\frac{1}{F}$ follows from C^0 and C^1 estimates. □

4.3 C^2 Estimates

Denote $\varphi \equiv \frac{1}{u}$, φ satisfies the following evolution equation.

Proposition 4.13. *Let ρ be a positive solution to (68) on $S^n \times [0, T)$. We have*

$$\partial_t\varphi = \frac{1}{F^2}\dot{F}^{ij}\nabla^i\nabla_j\varphi - \frac{\varphi}{F^2}\dot{F}(h^2) - \frac{2}{F^2\varphi}\dot{F}(\nabla\varphi, \nabla\varphi) + r\varphi + r\varphi^{-1}\nabla_l\varphi < X, \nabla_l X > . \qquad (80)$$

Proof. We first write down the evolution equation of u using (55), (58) and (44). We work on local orthonormal frames on $M(t)$.

$$
\begin{aligned}
u_t &= \langle X_t, \nu \rangle + \langle X, \nu_t \rangle \\
&= \frac{1}{F} - ru - \sum_l \langle X, X_l \rangle (\frac{1}{F} - ru)_l \\
&= \frac{1}{F} - ru + \sum_l \langle X, X_l \rangle (\frac{F^{ij} h_{ij,l}}{F^2} + ru_l) \\
&= \frac{1}{F} - ru + \frac{F^{ij}(u_{ij} - h_{ij} + (h^2)_{ij} u)}{F^2} + r \sum_l \langle X, X_l \rangle u_l \\
&= \frac{1}{F} - ru + \frac{F^{ij} u_{ij}}{F^2} - \frac{1}{F} + \frac{F^{ij}(h^2)_{ij} u}{F^2} + r \sum_l \langle X, X_l \rangle u_l
\end{aligned}
$$

Proposition follows from above identity by inserting $u = \frac{1}{\varphi}$. $\qquad\square$

Proposition 4.14. *Let ρ be a positive solution to (68) on $S^n \times [0, T)$. We have*

$$
\begin{aligned}
\partial_t(\varphi h_j^i) &= \frac{1}{F^2} \dot{F}^{kl} \nabla^k \nabla_l(\varphi h_j^i)) - \frac{2\varphi}{F^3} \nabla^i F \nabla_j F + \frac{\varphi}{F^2} \ddot{F}^{kl,mn} \nabla^i h_l^k \nabla_j h_n^m \\
&\quad - \frac{2}{F^2 \varphi} \dot{F}^{kl} \nabla^k \varphi \nabla_l(\varphi h_j^i) + r \nabla_l(\varphi h_j^i) < \nabla_l X, X > \\
&\quad - 2\varphi \left[\frac{(h^2)_j^i}{F} - r h_j^i \right].
\end{aligned}
\tag{81}
$$

Proof. Proof follows from (59) and Proposition 4.13. $\qquad\square$

Proposition 4.15. *Let ρ be a positive solution to (68) on $S^n \times [0, T)$ and let $\tilde{\kappa}(t) = \max_{x \in M_t^n} (\kappa_1(x), \cdots, \kappa_n(x))$. Then for $t > 0$,*

$$
\max_{M_t^n} \varphi \tilde{\kappa}(t) \le \max_{M_0^n} \varphi \tilde{\kappa}(0),
\tag{82}
$$

with the equality holds if and only if M_0 is a sphere centered at the origin. Since $\sigma_1(\kappa) > 0$, we have uniform curvature bounds.

Proof. Let x_0 be a point such that $h_1^1(x_0, t_0) = \kappa(t_0)$ for some direction e_1. By (81), and concavity of F,

$$
(\varphi h_1^1(x_0, t_0))_t \le -2\varphi \left[\frac{(h_1^1)^2}{F(\kappa)} - r h_1^1 \right].
$$

At x_0, $h_1^1 = \tilde{\kappa}(t) \ge \kappa_i$ for all i. By the monotonicity, homogeneity of F and by Lemma 4.5,

$$
\frac{h_1^1}{F(\kappa)} \ge \frac{1}{F(I)} \ge r.
\tag{83}
$$

We obtained at x_0, $(\varphi h_1^1(x_0, t_0))_t \le 0$.

We claim for any t_0, $(\varphi h_1^1(x_0, t))_t > 0$ unless $M(t_0)$ is the unit sphere centered at 0. Now suppose $(\varphi h_1^1(x_0, t))_t = 0$, all inequalities in (83) must be equalities. In particular,

$$r(t) = \frac{1}{F(I)}.$$

By Lemma 4.5 and normalization, $M(t_0)$ must be a sphere of radius 1. So $\kappa_1(x, t_0) = \cdots, \kappa_n(x, t_0) = 1, \forall x \in \mathbb{S}^n$ and we may use the standard spherical paramerization for $M(t_0)$. Suppose its center is $z \neq 0$, we may assume $z = (0, \cdots, 0, s)$ for some $-1 < s < 0$. Now

$$u(x, t_0) = 1 + s x_{n+1}, \quad \varphi(x, t_0) = \frac{1}{1 + s x_{n+1}}.$$

The minimum point is $x_0 = (0, \cdots, 0, 1)$, it follows from (81),

$$\partial_t (\varphi h_j^i) = \frac{1}{F^2} \dot{F}^{kl} \nabla^k \nabla_l (\varphi h_j^i) - 2\varphi \left[\frac{(h^2)_j^i}{F} - r h_j^i \right] = \frac{1}{F^2} \dot{F}^{kl} \nabla^k \nabla_l \varphi < 0,$$

contradiction.   □

We now prove Theorem 4.7.

Proof. By C^2 estimates and Proposition 4.12, $\kappa \in \Gamma_k$ is preserved along flow (68). By Lemma 2.7, the equation is uniform parabolic. We may apply the Krylov Theorem [31] and the standard parabolic theory to conclude the longtime existence and regularity for the flow. To get exponential convergence, we use the uniform ellipticity of F. There is $c_0 > 0$ independent of t,

$$\left(\frac{\partial F(B)}{\partial b_{ij}} \right)(x, t) \geq c_0 I, \quad \forall (x, t).$$

Thus, as $n \geq 2$,

$$\sum_i \frac{\partial F(B)}{\partial b_{ii}} \geq c_0 + \lambda_M \left(\frac{\partial F(B)}{\partial b_{ij}} \right),$$

where $\lambda_M(W)$ denoting the largest eigenvalue of W. By C^2 estimates, there is $\beta > 0$ independent of t such that

$$\frac{\omega^2}{F^2} \sum_{ij} \frac{\partial F(B)}{\partial b_{ij}} (\delta_{ij} |\overline{\nabla} \gamma|^2 - \gamma_i \gamma_j) \geq \beta |\overline{\nabla} \gamma|^2.$$

By Proposition 4.8,

$$\partial_t\Big(\frac{|\overline{\nabla}\gamma|^2}{2}\Big) \le \mathcal{L}_{lj}\overline{\nabla}_l\overline{\nabla}_j\Big(\frac{|\overline{\nabla}\gamma|^2}{2}\Big) + W_k\cdot\overline{\nabla}_k\Big(\frac{|\overline{\nabla}\gamma|^2}{2}\Big) - \beta|\overline{\nabla}\gamma|^2. \qquad (84)$$

Set $Q = e^{\beta t}\dfrac{|\overline{\nabla}\gamma|^2}{2}$, Q satisfies differential inequality

$$\partial_t Q \le \mathcal{L}_{lj}\overline{\nabla}_l\overline{\nabla}_j Q + W_k\cdot\overline{\nabla}_k Q. \qquad (85)$$

Therefore, Q is bounded from above independent of t. From there, we conclude $|\overline{\nabla}\gamma|^2 \to 0$ exponentially as $t \to \infty$. By our normalization, $\rho \to 1$ and $\overline{\nabla}\rho \to 0$ exponentially as $t \to \infty$.

For the exponential convergence of $\overline{\nabla}^m\rho$, apply integration by parts,

$$\int_{\mathbb{S}^n}|\overline{\nabla}^m\rho|^2 d\mu_{\mathbb{S}^n} \le C\Big(\int_{\mathbb{S}^n}|\overline{\nabla}^{m+1}\rho|^2 d\mu_{\mathbb{S}^n}\Big)^{\frac{1}{2}}\Big(\int_{\mathbb{S}^n}|\overline{\nabla}^{m-1}\rho|^2 d\mu_{\mathbb{S}^n}\Big)^{\frac{1}{2}}.$$

By the a priori estimates, $\|\overline{\nabla}^{m+1}\rho\|_{L^\infty(\mathbb{S}^n)} \le c_m$ for some c_m independent of t. An induction argument yields that, for each $m \in \mathbb{N}^+$, there is $C_m > 0, \beta_m > 0$, such that

$$\|\overline{\nabla}^m\rho\|_{L^2(\mathbb{S}^n)} \le C_m e^{-\beta_m t}.$$

The Sobolev Lemma implies $\overline{\nabla}^m\rho \to 0$ exponentially and $t \to \infty$, for each $m \in \mathbb{N}^+$. $\qquad\square$

We prove Theorem 4.1. In fact, the following is true.

Theorem 4.16. *Suppose Ω is a C^2 starshaped domain in $\mathbb{R}^{n+1}$. Assume $1 \le k \le n-1$, that*

$$\kappa(x) \in \overline{\Gamma}_k = \{\lambda \in \mathbb{R}^n | \sigma_l(\lambda) \ge 0, \forall l = 1, \cdots, k.\},$$

then the following inequality holds,

$$(V_{(n+1)-k}(\Omega))^{\frac{1}{n+1-k}} \le \mathbf{C}_{n,k}(V_{n-k}(\Omega))^{\frac{1}{n-k}}, \qquad (86)$$

where

$$\mathbf{C}_{n,k} = \frac{(V_{(n+1)-k}(B))^{\frac{1}{n+1-k}}}{(V_{n-k}(B))^{\frac{1}{n-k}}},$$

B is the standard ball in $\mathbb{R}^{n+1}$. The equality holds if and only if Ω is a ball.

Proof. Case 1. Ω is k-convex.

Inequality (86) follows directly from the above Proposition 4.6 and Theorem 4.7. We examine the equality case. Recall (66),

$$\partial_t \int_M \sigma_{k-1}(\kappa) d\mu_g = k \int_M \left[1 - \frac{\int_M \frac{\sigma_{k+1}(\kappa)\sigma_{k-1}(\kappa)}{\sigma_k(\kappa)} d\mu_g}{C_{n,k} \int_M \sigma_k(\kappa) d\mu_g} C_{n,k-1} \right] \sigma_{k-1} d\mu_g$$
$$\geq k \int_M \left[1 - \frac{\sigma_{k+1}(I)\sigma_{k-1}(I)}{\sigma_k^2(I)} \frac{C_{n,k-1}}{C_{n,k}} \right] \sigma_{k-1} d\mu_g = 0. \tag{87}$$

At any time $t_0 \geq 0$, inequality is strict in (87) unless

$$\frac{\sigma_{k+1}(\kappa)\sigma_{k-1}(\kappa)}{\sigma_{k+1}(I)\sigma_{k-1}(I)\sigma_k(\kappa)} = \frac{\sigma_k(\kappa)}{\sigma_k^2(I)}, \quad \text{a.e. in } M(t_0).$$

That is the equality is the case in (13), this implies $M(t_0)$ is umbilical almost everywhere. As $M(t_0)$ is C^2, it is umbilical everywhere. $M(t_0)$ is a round sphere for each $t \geq t_0$. In particular, if equality is held in (86), then M is a sphere.
Case 2. General case.

We may approximate Ω by k-convex starshaped domains. The inequality follows from the approximation. We now treat the equality case. We first note that both $\int_M \sigma_k d\mu_g$ and $\int_M \sigma_{k-1} d\mu_g$ are positive, since there exists at least one elliptic point on an embedded compact hypersurface in Euclidean space and also the k-convexity condition. Suppose Ω is a weakly k-convex starshaped domain with equality in (86) attained. Let $M_+ = \{x \in M \,|\, \sigma_k(\kappa(x)) > 0\}$. M_+ is open and nonempty since M is compact and embedded in $\mathbb{R}^{n+1}$. We claim that M_+ is closed. This would imply $M = M_+$, so Ω is k-convex, by *Case 1*, we may conclude Ω is a standard ball.

We now prove that M_+ is closed. Pick any $\eta \in C_0^2(M_+)$ compactly supported in M_+. Let M_s be the hypersurface determined by position function $X_s = X + s\eta\nu$, where X is the support function of M and ν is the unit outernormal of M at X. Let Ω_s be the domain enclosed by M_s. It is easy to show M_s is k-convex starshaped when s is small enough. Define

$$\mathcal{I}_k(\Omega) = \frac{V_{(n+1)-k}^{\frac{1}{n+1-k}}(\Omega)}{V_{n-k}^{\frac{1}{n-k}}(\Omega)}. \tag{88}$$

Therefore $\mathcal{I}_k(\Omega_s) - \mathcal{I}_k(\Omega) \leq 0$ for s small, i.e.

$$\frac{d}{ds}\mathcal{I}_k(\Omega_s)|_{s=0} = 0.$$

Simple calculation yields

$$\frac{d}{ds} \int_{M_s} \sigma_l(\kappa_s) d\mu_{g_s}|_{s=0} = (l+1) \int_M \sigma_{l+1}(\kappa)\eta d\mu_g.$$

Therefore,

$$\frac{d}{ds}\mathcal{I}_k(\Omega_s)|_{s=0} = A \int_M (\sigma_{k+1}(\kappa) - c_1\sigma_k(\kappa))\eta d\mu_g = 0,$$

for some constant $A > 0$ with $c_1 = \frac{k(n-k)}{(k+1)(n-k+1)} \frac{1}{\mathcal{I}(B)^{n-k+1}(\int_M \sigma_k)^{\frac{1}{n-k}}} > 0$ and for all $\eta \in C_0^2(M_+)$. Thus,

$$\sigma_{k+1}(\kappa(x)) = c_1\sigma_k(\kappa(x)), \qquad \forall x \in M_+. \tag{89}$$

It follows from the Newton-MacLaurine inequality, there is a dimensional constant $\tilde{C}_{k,n}$ such that

$$\sigma_{k+1}(\kappa(x)) \leq \tilde{C}_{k,n}\sigma_k^{1+1/k}(\kappa(x)), \quad \forall x \in M_+.$$

In view of (89), there is a positive constant c_2, such that

$$\sigma_k(\kappa(x)) \geq c_2 > 0, \qquad \forall x \in M_+, \tag{90}$$

where $c_2 = (\frac{c_1}{\tilde{C}_{k,n}})^k$ is a positive constant depending only on n, k, and Ω. (90) implies M_+ is closed. $\qquad\square$

5 Appendix

We present Garding's theory of hyperbolic polynomials here.

Definition 5.1. Let P be a homogeneous polynomial of degree m in a finite vector space V. For $\theta \in V$, P is called hyperbolic at θ if $P(\theta) \neq 0$ and the equation $P(x + t\theta) = 0$ (as a polynomial of $t \in \mathbb{C}$) has only real roots for every $x \in V$. We say P is complete if $P(x + ty) = P(x)$ for all x, t implies $y = 0$.

Proposition 5.2. *Suppose P is hyperbolic at θ, then the component Γ of θ in $\{x \in V; P(x) \neq 0\}$ is a convex cone, the zeros of $P(x + ty)$ as a polynomial in t are real $\forall x, y \in V$. The polynomial $\frac{P(x)}{P(\theta)}$ is real, and it is positive when $x \in \Gamma$. Furthermore, $(\frac{P(x)}{P(\theta)})^{\frac{1}{m}}$ is concave and homogeneous of degree 1 in Γ, equal to 0 on the boundary of Γ.*

Proof. We normalize $P(\theta) = 1$, then there exist $t_j \in \mathbb{R}$, $j = 1, \cdots, m$, such that

$$P(x + t\theta) = (t - t_1) \times \ldots \times (t - t_m).$$

In particular, $P(x) = (-t_1) \times \ldots \times (-t_m) \in \mathbb{R}$. Set

$$\Gamma_\theta = \{x \in V; P(x + t\theta) \neq 0, t \geq 0\}.$$

Γ_θ is open and $\theta \in \Gamma_\theta$ as $P(\theta + t\theta) = (1 + t)^m P(\theta)$ only has the zero $t = -1$. Notice that Γ_θ is also closed in $\{x \in V; P(x) \neq 0\}$. If $x \in \bar{\Gamma}_\theta$, then $P(x + t\theta) \neq 0$, when $t > 0$. Hence,

$$\Gamma_\theta = \{x \in \bar{\Gamma}_\theta, P(x) \neq 0\}.$$

If $x \in \Gamma_\theta$, then $x + t\theta \in \Gamma_\theta$ when $t > 0$. This implies that Γ_θ is connected, Therefore $\lambda x + \mu \theta \in \Gamma_\theta$ for all $\lambda > 0, \mu > 0$. That is, Γ_θ is star-shaped with respect to θ and $\Gamma_\theta = \Gamma$.

For $y \in \Gamma$ and $\delta > 0$ fixed,

$$E_{y,\delta} = \{x \in V; P(x + i\delta\theta + isy) \neq 0, Re(s) \geq 0\}$$

is open. If $s \neq 0$, $P(i\delta + isy) = (is)^m P(\frac{\delta\theta}{s} + y) = 0$, the hyperbolicity implies $s < 0$. That is, $0 \in E_{y,\delta}$. If $x \in \bar{E}_{y,\delta}$ and $Res > 0$, then Hurwitz' theorem implies $P(x + i\delta\theta + isy) \neq 0$. This is still true when $Re(s) = 0$ since $x + isy$ is real. Therefore, $E_{y,\delta}$ is both open and closed, and $E_{y,\delta} = V$. Thus,

$$P(x + i(\delta\theta + y)) \neq 0, \forall x \in \mathbb{R}^n, y \in \Gamma, \delta > 0.$$

For Γ is open, the above remains true for $\delta = 0$. Equation $P(x + ty) = 0$ has only real roots, for if $t = t_1 + it_2$ is a root with $t_2 \neq 0$ we would get $P(\frac{x+t_1y}{t_2} + iy) = 0$. This means that y can play the role of θ, Γ is star-shaped with respect to every point in Γ. The convexity of Γ follows. We also have $P(y) > 0$ for all $y \in \Gamma$.

We now prove the concavity statement in the proposition. As $P(x + ty)$ has only real roots for $y \in \Gamma$, there are $t_j \in \mathbb{R}$, $j = 1, \ldots, m$,

$$P(x + ty) = P(y)(t - t_1) \times \ldots \times (t - t_m).$$

In turn,

$$P(sx + y) = P(y)(1 - st_1) \times \ldots (1 - st_m).$$

If $sx + y \in \Gamma$, we must have $1 - st_j > 0$ for every j. If $f(s) = \log P(sx + y)$, then

$$f'(s) = -\sum \frac{t_j}{1 - st_j}, \quad f''(s) = -\sum \frac{t_j^2}{(1 - st_j)^2}.$$

Therefore, by Cauchy-Schwarz inequality,

$$m^2 e^{-\frac{f(s)}{m}} \frac{d^2(e^{\frac{f(s)}{m}})}{ds^2} = f'(s)^2 + m f''(s)$$

$$= (\sum \frac{t_j}{1 - st_j})^2 - m \sum \frac{t_j^2}{(1 - st_j)^2} \leq 0.$$

$\square$

If P is a homogeneous polynomial of degree m. For $x^l = (x_1^l, \ldots, x_n^l) \in V$, $l = 1, \ldots, m$, we denote $< x^l, \frac{\partial}{\partial x} >= \sum_1^n x_j^l \frac{\partial}{\partial x_j}$ as a vector field. We define the complete polarization of P as

$$\tilde{P}(x^1, \ldots, x^m) = \frac{1}{m!} < x^1, \frac{\partial}{\partial x} > \ldots < x^m, \frac{\partial}{\partial x} > P(x).$$

It is a multilinear and symmetric in $x^1, \ldots, x^m \in V$, independent of x, and that

$$\tilde{P}(x, \ldots, x) = \frac{1}{m!} \frac{d^m}{dt^m} P(tx) = P(x), \forall x \in V.$$

And

$$P(t_1 x^1 + \ldots + t_m x^m) = m! t_1 \ldots t_m \tilde{P}(x^1, \ldots, x^m) + \ldots$$

where the dots denote terms not containing all the factors t_j.

Lemma 5.3. *If P is hyperbolic at θ and $m > 1$, then for any $y = (y_1, \ldots, y_n) \in \Gamma$,*

$$Q(x) = \sum_1^n y_j \frac{\partial}{\partial x_j} P(x)$$

is also hyperbolic at θ. In general, if $x^1, \ldots, x^l \in \Gamma$ for some $l < m$, then

$$\tilde{Q}_l(x) = \tilde{P}(x^1, \ldots, x^l, x, \ldots, x)$$

is hyperbolic at θ.

The proof is immediate by Rolle's theorem. Using polarization and Lemma 5.3, we list some of important examples of hyperbolic polynomials.

Corollary 5.4. *The following polynomials are hyperbolic.*

1. *The polynomial $P = (x_1)^2 - (x_2)^2 - \ldots - (x_n)^2$ is hyperbolic at $(1, 0, \ldots, 0)$.*
2. *The polynomial $P = x_1 \ldots x_n$ is complete hyperbolic at any θ with $P(\theta) \neq 0$. The positive cone Γ of P at $(1, \ldots, 1)$ is*

$$\Gamma = \{x = (x_1, \ldots, x_n); x_j > 0, \quad \forall j\}.$$

3. *In general the elementary symmetric function $\sigma_k(x)$ is complete hyperbolic at $(1, \ldots, 1)$, the corresponding positive cone Γ_k is*

$$\Gamma_k = \{\sigma_l(x) > 0, \forall l \leq k\}.$$

4. *Let $\mathcal{S}$ denote set of all real $n \times n$ symmetric matrices. Then $\sigma_k(W), W \in \mathcal{S}$ is complete hyperbolic at the identity matrix, the corresponding positive cone is*

$$\Gamma_k = \{\sigma_l(W) > 0, \forall l \leq k\}.$$

5. *For $W^1, \ldots, W^l \in \Gamma_k, l < k$, then $Q_l(W) = \tilde{P}(W^1, \ldots, W^l, W, \ldots, W)$ is complete hyperbolic in Γ_k.*

Lemma 5.5. *Suppose P is a second order complete hyperbolic polynomial. Suppose both roots of $f(s) = P(sy + w)$ vanishing for some $y \in \Gamma$ and $w \in V$. Then, all the roots of $g(s) = P(sz + w)$ are vanishing for any $z \in \Gamma$.*

Proof. Since $P(y + tw) = P(y) \neq 0$ for all t, we must have $y + tw \in \Gamma$. By the convexity of Γ, we have $z + tw \in \Gamma$ for all t. So, $P(z + tw) \neq 0$. For any $z \in \Gamma$ and all t,

$$P(z)(1 + t\lambda_1)(1 + t\lambda_2) = P(z + tw) \neq 0,$$

λ_1, λ_2 are the roots of $P(sz + w)$. Since t is arbitrary, this gives $\lambda_1 = \lambda_2 = 0$. □

Lemma 2.4 is a special case of the following proposition.

Proposition 5.6. *Suppose P a homogenous polynomial of degree m, suppose it is hyperbolic at θ and $P(\theta) > 0$, then $\forall x^1, \ldots, x^m \in \Gamma$,*

$$P^2(x^1, x^2, x^3, \cdots, x^m) \geq P(x^1, x^1, x^3, \cdots, x^m)P(x^2, x^2, x^3, \cdots, x^m)$$

$$P(x^1, \ldots, x^m) \geq P(x^1)^{\frac{1}{m}} \ldots P(x^m)^{\frac{1}{m}}. \tag{91}$$

If P is complete, the equality holds if and only if all x^j are pairwise proportional. This is also equivalent that for $x, y \in \Gamma$ not proportional, the function $h(t) = P(x + ty)^{\frac{1}{m}}$ is strictly concave in $t > 0$. If P is complete, then $\tilde{Q}_l(X) = \tilde{P}(x^1, \ldots, x^l, x, \ldots, x)$ is complete if $m - l \geq 2$ and $x^1, \ldots, x^l \in \Gamma$. In particular, $\tilde{P}(x^1, \ldots, x^m) > 0$ if $x^1 \in \bar{\Gamma}$ and $x^j \in \Gamma$ when $m \geq 2$.

Proof. Since $P^{\frac{1}{m}}(X)$ is concave in Γ, it follows that for any $x, y \in \Gamma$, $h(t) = P(x + ty)^{\frac{1}{m}}$ is concave in $t > 0$. So, $h''(t) \leq 0$. A direct computation yields

$$h''(0) = (m-1)(\tilde{P}(y, y, x, \ldots, x)P(X) - \tilde{P}(y, x, \ldots, x)^2)P(x)^{\frac{1}{m}-2}.$$

We get the inequality

$$\tilde{P}(y, y, x, \ldots, x) P(X) \le \tilde{P}(y, x, \ldots, x)^2.$$

In turn, it implies

$$\tilde{P}(y, x, \ldots, x)^m \ge P(y) P(x)^{m-1}.$$

We now apply induction argument. Take $y = x^1$ and assuming that (91) is already proved for hyperbolic polynomials of degree $m - 1$. Let $Q(x) = \tilde{P}(y, x, \ldots, x)$, we get

$$\tilde{P}(x^1, \ldots, x^m) \ge (Q(x^2) \ldots Q(x^m))^{\frac{1}{(m-1)}}$$

$$\ge (P(x^1) P(x^2)^{m-1} \ldots P(x^1) P(x^m)^{m-1})^{\frac{1}{m(m-1)}},$$

which proves (91).

To prove the last statement in the proposition, it suffices to show that if $m \ge 3$, Q (defined above) is complete. suppose $Q(x) = Q(x + tz)$ for all x, t. In particular, $Q(y + tz) = Q(y)$. That means that $Q(ty + z) = Q(ty)$, so $P(ty + z) - P(ty) = a$ is independent of t. Since the zeros of $P(ty) + a = t^m P(y) + a$ must all be real, it follows that $a = 0$. This $P(y + sz) = P(y) \ne 0$ for all s, so it follows that $y + sz \in \Gamma$. Hence,

$$\frac{(sx + y + sz)}{(s + 1)} \in \Gamma, \forall x \in \Gamma, s > 0.$$

Letting $s \to \infty$, we conclude that $x + z \in \bar{\Gamma}$ for all $x \in \Gamma$. This implies $x + z \in \Gamma$. We can replace z by tz for any t, so $x + tz \in \Gamma$ for all t and $x \in \Gamma$. Thus $P(z + sx)$ can not have any zeros $\ne 0$, so $P(z + sx) = s^m P(x)$. That is $P(x + tz) = P(x)$ for all t and all $x \in \Gamma$. Since P is analytic, that means $P(x + tz) = P(x)$ for all t and all $x \in V$. By the completeness assumption on P, $z = 0$.

Finally, we discuss the equality case in (91). By the above, we may assume $m = 2$. If the equality holds, we have $P(y) P(x) = \tilde{P}(y, x)^2$. This implies the roots of the second order polynomial $p(t) = P(x + ty)$ are equal, i.e., $t_1 = t_2 = -\lambda \ne 0$. In turn, for all t,

$$P(y + (t + \lambda)^{-1}(x - \lambda y)) = (t + \lambda)^{-2} P(ty + x) = P(y).$$

That is both roots of the polynomial $f(s) = P(sy + (x - \lambda y))$ are vanishing.

From Lemma 5.5, we have $P(z + t(x - \lambda y)) = P(z)$ for all $z \in \Gamma$ and all t. Since Γ is open and P is analytic, $P(z + t(x - \lambda y)) = P(z)$ for all z and all t. By the completeness of P, $x - \lambda y = 0$. That is, x and y are proportional. $\square$

6   Notes

1. The definition of curvature measures in this notes follows from Federer [12], where he used Steiner's formula to define them for sets of positive reach. Alexandrov [3] initiated the problem of prescribing curvature measure $\mathcal{C}_0$, which he called the integral curvature. The problem of prescribing 0-th curvature measure is often referred as the Alexandrov problem in literature. It was Alexandrov who formulated the problem through radial parametrization. The existence and uniqueness of solutions were obtained by A.D. Alexandrov [3]. It can be deduced to a Monge-Ampére type equation on $\mathbb{S}^n$. For $n = 2$ the regularity of solutions of the Alexandrov problem in the elliptic case was proved by Pogorelov [37] and for higher dimensional cases, it was solved by Oliker [35]. The general regularity results (degenerate case) of the problem were obtained in [20]. The problem of prescribing general k-th curvature measures was settled for starshaped hypersurfaces recently in [27], though C^0 and C^1 estimates were obtained in [19] some time ago. The proof of Lemma 3.4 presented here is due to Junfang Li (Li, private notes (2012)), which can apply to more general curvature equations. Another proof of gradient estimate for (26) appeared in [25], there the question of when solution to (26) is discussed.

2. The presentation of theory of hyperbolic polynomials in Appendix basically follows the original paper of Garding [14]. Caffarelli-Nirenberg-Spruck [5] developed the study of k-Hessian equation in the category of Γ_k, followed by [6] for k-curvature equation. The proof of Lemma 2.7 is from [30], which in turn is inspired by Marcus and Lopes [32]. Lemma 2.8 was proved in [27]. Using $\frac{\kappa}{u}$ in C^2 estimates for k-curvature equation on star-shaped hypersurfaces was introduced in [6]. The complication for (26) is that the right hand side depends on $\nabla\rho$, the standard concavity of $\sigma_k^{\frac{1}{k}}$ is not sufficient in this case. C^2 estimate is still open for k-curvature equation on star-shaped k-convex hypersurfaces with general right hand side

$$\sigma_k(\kappa) = f(\nabla\rho(x), \rho(x), x), \quad x \in \mathbb{S}^n.$$

In a recent work [26] established C^2 estimates for admissible solutions of above equation in the case $k = 2$ and for convex solutions for general k.

3. The classical isoperimetric inequalities for quermassintegrals of convex bodies are the consequence of the Alexandrov-Fenchel inequality [1, 2] in convex geometry. Trudinger was the first to consider such inequalities for k-convex domains in [40]. Theorem 4.1 was proved in [17]. The proof in [17] used un-normalized inverse mean curvature type flow for starshaped hypersurfaces studied by Gerhardt [15] and Urbas [41], where they established longtime existence and exponential convergence for a class of more general type of inverse mean curvature flow. In Sect. 3, we use normalized flow (63), which was initially devised in (Guan and Li, private notes) when they did not realize that the work of [15,41] would imply the monotonicity of the isoperimetric ratio $\mathcal{I}_k$ in (88). Flow

(63) considered here has an advantage that one can see how to design a flow to fit the monotonicity. Similar design was used previously in conformal geometry in [22,23]. Junfang Li pointed out that, one may also pick $r(t) \equiv \frac{1}{F(I)}$ in (63), as in a recent paper [18]. With this choice of r, the proof of C^0 estimates for flow (63) can be simplified. The monotonicity in Proposition 4.6 is reversed as

$$\int_M \sigma_k d\mu_g \text{ is monotonically non-increasing; } \int_M \sigma_{k-1} d\mu_g \text{ is a constant.}$$

It is an open question if (54) is valid without the starshapedness condition. In the case $k = 1$, Huisken [29] verified the inequality replacing the star shapedness by the assumption that $\partial\Omega$ is outward-minimizing. Again, in the case $k = 1$, (54) was proved for general 1-convex domains in [7] for some constant $\mathbf{c}$ which is a not sharp. Under additional condition that Ω is $k + 1$-convex (without starshapedness assumption), inequality (54) is proved in [8] with some no-sharp constant $\mathbf{c}$.

4. The normalized inverse mean curvature flow

$$X_t = (\frac{1}{H} - \frac{u}{n})v \qquad (92)$$

preserves the surface area and increases the enclosed volume. This implies the isoperimetric inequality for mean convex star-shaped domain. The statement can be checked as below.

$$\begin{aligned} \frac{d}{dt}\int_M d\mu_g &= \int_M (\frac{1}{H} - \frac{u}{n})H d\mu \\ &= \frac{1}{n}\int_M (n - uH)d\mu \\ &= 0. \end{aligned} \qquad (93)$$

The evolution of the volume $V(t)$ is

$$\begin{aligned} \frac{d}{dt}V &= \int_M (\frac{1}{H} - \frac{u}{n})d\mu \\ &= \int_M \frac{1}{H}d\mu - \frac{n+1}{n}V \\ &\geq 0. \end{aligned} \qquad (94)$$

where the last inequality comes from an inequality proved by Ros in [39], see formula (5) on page 449.

5. The prescribing measure problem is a counter part of the Christoffel-Minkowski problem, which is the problem of prescribing area measures for convex bodies. The Minkowski problem was considered by Minkowski in [33] in 1897. The differential geometric setting of the problem was solved in early 1950s by

Nirenberg [34] and Pogorelov [36] for $n = 2$. The solution of the Minkowski problem in higher dimension came much later in 1970s by Cheng-Yau [9] and Pogorelov [38]. The Minkowski problem is a special case ($k = n$) of the problem of prescribing general k-th ($1 \leq k \leq n$) area measures in convex geometry. At the other end ($k = 1$), it is the Christoffel problem. This case has been settled completely by Firey [13]. In general, the problem of prescribing k-th is termed the Christoffel-Minkowski problem. It is equivalent to solve the following equation

$$\sigma_k(u_{ij} + u\delta_{ij}) = \varphi \quad \text{on} \quad \mathbb{S}^n, \tag{95}$$

with convexity requirement $(u_{ij} + u\delta_{ij}) > 0$.

The intermediate Christoffel-Minkowski problem ($1 < k < n$) is still open, except for some special cases. There are also some sufficient conditions, we refer to [38] and [21]. The necessary and sufficient condition for the existence of admissible solutions of (95) is known (e.g., [24]). The main difficulty lies in the question of convexity for the admissible solutions (which in general are not *convex*) of (95).

6. The Minkowski problem can also be considered as a problem of prescribing the Gauss curvature on outernormals of convex hypersurfaces. The similar question was raised for other Weingarten curvature functions $\sigma_k(\kappa_1, \cdots, \kappa_n)$ for fixed $1 \leq k \leq n$ in [4] and [10]. The corresponding equation is

$$\frac{\sigma_n}{\sigma_{n-k}}(u_{ij} + u\delta_{ij}) = f \quad \text{on} \quad \mathbb{S}^n. \tag{96}$$

When $1 \leq k < n$, very little is known for this problem. No uniqueness result is known except the case $n = 2$ (e.g., see [4]). If the prescribed curvature function is invariant under an automorphic group G without fixed points, the problem is solvable [16].

Acknowledgements Large part of the material in this lecture notes are based on joint works with Junfang Li [17,27]. I would like to thank him for many helpful discussions and valuable comments regarding the exposition of the notes. Research of the first author was supported in part by an NSERC Discovery Grant.

References

1. A.D. Alexandrov, Zur Theorie der gemischten Volumina von konvexen korpern, II. Neue Ungleichungen zwischen den gemischten Volumina und ihre Anwendungen (in Russian). Mat. Sbornik N.S. **2**, 1205–1238 (1937)
2. A.D. Alexandrov, Zur Theorie der gemischten Volumina von konvexen korpern, III. Die Erweiterung zweeier Lehrsatze Minkowskis uber die konvexen polyeder auf beliebige konvexe Flachen (in Russian). Mat. Sbornik N.S. **3**, 27–46 (1938)

3. A.D. Alexandrov, Existence and uniqueness of a convex surface with a given integral curvature. Doklady Akademii Nauk Kasah SSSR **36**, 131–134 (1942)
4. A.D. Alexandrov, Uniqueness theorems for surfaces in the large I. Vestnik Leningrad Univ. **11**, 5–17 (1956); Am. Soc. Trans. Ser. 2 **21**, 341–354 (1962)
5. L. Caffarelli, L. Nirenberg, J. Spruck, The Dirichlet problem for nonlinear second order elliptic equations, III: functions of the eigenvalues of the Hessian. Acta Math. **155**, 261–301 (1985)
6. L.A. Caffarelli, L. Nirenberg, J. Spruck, *Nonlinear Second Order Elliptic Equations IV: Starshaped Compact Weigarten Hypersurfaces*, ed. by Y. Ohya, K. Kasahara, N. Shimakura. Current Topics in Partial Differential Equations (Kinokunize, Tokyo, 1985), pp. 1–26
7. P. Castillon, Submanifolds, isoperimetric inequalities and optimal transportation. J. Funct. Anal. **259**, 79–103 (2010)
8. A. Chang, Y. Wang, *The Aleksandrov-Fenchel Inequalities for Quermassintegrals on $k + 1$-Convex Domains*. Preprint (2011)
9. S.Y. Cheng, S.T. Yau, On the regularity of the solution of the n-dimensional Minkowski problem. Comm. Pure Appl. Math. **24**, 495–516 (1976)
10. S.S. Chern, Integral formulas for hypersurfaces in Euclidean space and their applications to uniqueness theorems. J. Math. Mech. **8**, 947–955 (1959)
11. L.C. Evans, Classical solutions of fully nonlinear, convex, second order elliptic equations. Comm. Pure Appl. Math. **35**, 333–363 (1982)
12. H. Federer, Curvature measures. Trans. Am. Math. Soc. **93**, 418–491 (1959)
13. W.J. Firey, The determination of convex bodies from their mean radius of curvature functions. Mathematik **14**, 1–14 (1967)
14. L. Garding, An inequality for hyperbolic polynomials. J. Math. Mech **8**, 957–965 (1959)
15. C. Gerhardt, Flow of nonconvex hypersurfaces into spheres. J. Differ. Geom. **32**, 299–314 (1990)
16. B. Guan, P. Guan, Hypersurfaces with prescribed curvatures. Ann. Math. **156**, 655–674 (2002)
17. P. Guan, J. Li, The quermassintegral inequalities for k-convex starshaped domains. Adv. Math. **221**, 1725–1732 (2009)
18. P. Guan, J. Li, *A Mean Curvature Type Flow in Space Forms*. Preprint (2012)
19. P. Guan, Y.Y. Li, *Unpublished Research Notes* (1995)
20. P. Guan, Y.Y. Li, $C^{1,1}$ estimates for solutions of a problem of Alexandrov. Comm. Pure Appl. Math. **50**, 189–811 (1997)
21. P. Guan, X. Ma, The Christoffel-Minkowski problem I: convexity of solutions of a Hessian equation. Inventiones Mathematicae **151**, 553–577 (2003)
22. P. Guan, G. Wang, A fully nonlinear conformal flow on locally conformally flat manifolds. Journal fur die reine und angewandte Mathematik **557**, 219–238 (2003)
23. P. Guan, G. Wang, Geometric inequalities on locally conformally flat manifolds. Duke Math. J. **124**, 177–212 (2004)
24. P. Guan, X. Ma, F. Zhou, The Christoffel-Minkowski problem III: existence and convexity of admissible solutions. Comm. Pure Appl. Math. **59**, 1352–1376 (2006)
25. P. Guan, C.S. Lin, X. Ma, The existence of convex body with prescribed curvature measures. Int. Math. Res. Not. **2009**, 1947–1975 (2009)
26. P. Guan, C. Ren, Z. Wang, *Global Curvature Estimates for Convex Solutions of σ_k-Curvature Equations*. Preprint (2012)
27. P. Guan, J. Li, Y.Y. Li, Hypersurfaces of prescribed curvature measures. Duke Math. J. **161**, 1927–1942 (2012)
28. G. Hardy, J. Littlewood, G. Polya, *Inequalities* (Cambridge University Press, London, 1952)
29. G. Huisken, in preparation
30. G. Huisken, C. Sinestrari, Convexity estimates for mean curvature flow and singularities of mean convex surfaces. Acta. Math. **183**, 45–70 (1999)
31. N.V. Krylov, Boundely inhomogeneous elliptic and parabolic equations in domains. Izvestin Akad. Nauk. SSSR **47**, 75–108 (1983)
32. M. Marcus, L. Lopes, Inequalities for symmetric functions and Hermitian matrices. Can. J. Math. **9**, 305–312 (1957)

33. H. Minkowski, Allgemeine Lehrsätze über die konvexen Polyeder. Nachr. Ges. Wiss. Gottin-gen, Mathematisch-Physikalishe Klass, Zeitschriftenband, Heft 2, 198–219 (1897)
34. L. Nirenberg, The Weyl and Minkowski problems in differential geometry in the large. Comm. Pure Appl. Math. **6**, 337–394 (1953)
35. V.I. Oliker, Existence and uniqueness of convex hypersurfaces with prescribed Gaussian curvature in spaces of constant curvature. Sem. Inst. Mate. Appl. "Giovanni Sansone", Univ. Studi Firenze, 1–39 (1983)
36. A.V. Pogorelov, Regularity of a convex surface with given Gaussian curvature. Mat. Sb. **31**, 88–103 (1952)
37. A.V. Pogorelov, *Extrinsic Geometry of Convex Surfaces* (Nauka, Moscow, 1969); English transl., Transl. Math. Mono., vol. 35 (American Mathematical Society, Providence, 1973)
38. A.V. Pogorelov, *The Minkowski Multidimensional Problem* (Wiley, New York, 1978)
39. A. Ros, Compact hypersurfaces with constant higher order mean curvatures. Rev. Mat. Iberoamericana **3**(3–4), 447–453 (1987)
40. N. Trudinger, Isoperimetric inequalities for quermassintegrals. Ann. Inst. H. Poincarè Anal. Non Linèaire **11**, 411–425 (1994)
41. J. Urbas, On the expansion of starshaped hypersurfaces by symmetrics functions of their principal curvatures. Mathmatische Zeitschrift **205**, 355–372 (1990)

Refraction Problems in Geometric Optics

Cristian E. Gutiérrez

Abstract We present a description of the far field and the near field problems for refraction when the source of energy is located at one point. The far field problem is solved using mass transportation and also a variant of the Minkowski method. Maxwell equations are developed and the boundary conditions studied to obtain Fresnel formulas. These are used to present a model for refraction that takes into consideration the energy used in internal reflection.

1 Introduction

These are notes expanding the material of a mini course I taught on Monge-Ampère type equations and geometric optics in June 2012 at Cetraro. The purpose of these lectures was to explain basic facts about the Monge-Ampère equation and the application of these ideas to pose and solve problems in geometric optics concerning refraction with prescribed input and output energies. These problems are basically of two types: the far field and the near field. In the far field case the goal is to send radiation into a set of directions and in the second is to send radiation to a specific target set. The far field case can be treated with optimal transportation methods, Sect. 3, but the near field does not. Instead in this case, a method based on the Minkowski method in convex geometry can be used. In fact, with this method one can treat both far and near fields problems. The method is illustrated in Sect. 4 in the far field case. Refraction and reflection occur simultaneously, in other words, if an incident ray is refracted there is always a percentage of the incident ray that is internally reflected. The fractions of energy refracted and reflected are given by the Fresnel formulas. Beginning with Maxwell's equations this is developed and

C.E. Gutiérrez (✉)
Department of Mathematics, Temple University, Philadelphia, PA 19122, USA
e-mail: gutierre@temple.edu

L. Capogna et al., *Fully Nonlinear PDEs in Real and Complex Geometry and Optics*,
Lecture Notes in Mathematics 2087, DOI 10.1007/978-3-319-00942-1_3,
© Springer International Publishing Switzerland 2014

explained in Sect. 5. The application to the refraction problem with loss of energy is
described in Sect. 5.13.

These notes include the some of my work with Qingbo Huang and Henok Mawi,
and the relevant references are [7–9]. The equations describing the solutions to these
problems are Monge-Ampère type equations whose derivation is included in [7],
and [9].

The notes are organized as follows. Section 2 contains the Snell law of refraction
in vector form. This is simple but essential for the development of the results. In
particular, we introduce in this section the notion of surfaces having the uniform
refraction property used later to give the definition of refractors. Section 3 contains
basic facts about optimal mass transport and its application to solve the refractor
problem in the far field case. Section 4 introduces a method to solve the far field
refractor problem that has its roots in the Minkowski method. Section 5 contains a
detailed development of Maxwell's equations, the so called boundary conditions
explaining the propagation across two different materials, and its application to
deduce the Fresnel formulas. These formulas are written in a convenient form that
is used finally to propose, in Sect. 5.13, a model for the refractor problem with loss
of energy in internal reflection.

It is a pleasure to thank the CIME for their support in the organization of this
series of lectures and in particular, to Elvira Mascolo and Pietro Zecca for their
great support and help that made this series possible.

2   Snell's Law of Refraction

2.1   In Vector Form

Suppose Γ is a surface in $\mathbb{R}^3$ that separates two media I and II that are homogeneous
and isotropic. Let v_1 and v_2 be the velocities of propagation of light in the media I
and II respectively. The index of refraction of the medium I is by definition $n_1 =
c/v_1$, where c is the velocity of propagation of light in the vacuum, and similarly
$n_2 = c/v_2$. If a ray of light[1] having direction $x \in S^2$ and traveling through the
medium I hits Γ at the point P, then this ray is refracted in the direction $m \in S^2$
through the medium II according with the Snell law in vector form:

$$n_1(x \times v) = n_2(m \times v), \tag{1}$$

where v is the unit normal to the surface to Γ at P going towards the medium II.
The derivation of this formula is in Sect. 2.2. This has several consequences:

---

[1]Since the refraction angle depends on the frequency of the radiation, we assume radiation is
monochromatic.

(a) the vectors x, m, ν are all on the same plane (called plane of incidence);

(b) the well known Snell law in scalar form

$$n_1 \sin \theta_1 = n_2 \sin \theta_2,$$

where θ_1 is the angle between x and ν (the angle of incidence), θ_2 the angle between m and ν (the angle of refraction),

From (1), $(n_1 x - n_2 m) \times \nu = 0$, which means that the vector $n_1 x - n_2 m$ is parallel to the normal vector ν. If we set $\kappa = n_2/n_1$, then

$$x - \kappa m = \lambda \nu, \tag{2}$$

for some $\lambda \in \mathbb{R}$. Taking dot products we get $\lambda = \cos \theta_1 - \kappa \cos \theta_2$, $\cos \theta_1 = x \cdot \nu > 0$, and $\cos \theta_2 = m \cdot \nu = \sqrt{1 - \kappa^{-2}[1 - (x \cdot \nu)^2]}$, so

$$\lambda = x \cdot \nu - \kappa \sqrt{1 - \kappa^{-2}(1 - (x \cdot \nu)^2)}. \tag{3}$$

Refraction behaves differently for $\kappa < 1$ and for $\kappa > 1$.

2.1.1 $\kappa < 1$

This means $v_1 < v_2$, and so waves propagate in medium II faster than in medium I, or equivalently, medium I is denser than medium II. In this case the refracted rays tend to bent away from the normal, that is the case for example, when medium I is glass and medium II is air. Indeed, from the scalar Snell law, $\sin \theta_1 = \kappa \sin \theta_2 < \sin \theta_2$ and so $\theta_1 < \theta_2$. For this reason, the maximum angle of refraction θ_2 is $\pi/2$ which, from the Snell law in scalar form, is achieved when $\sin \theta_1 = n_2/n_1 = \kappa$. So there cannot be refraction when the incidence angle θ_1 is beyond this critical value, that is, we must have $0 \leq \theta_1 \leq \theta_c = \arcsin \kappa$.[2] Once again from the Snell law in scalar form,

$$\theta_2 - \theta_1 = \arcsin(\kappa^{-1} \sin \theta_1) - \theta_1 \tag{4}$$

and it is easy to verify that this quantity is strictly increasing for $\theta_1 \in [0, \theta_c]$, and therefore $0 \leq \theta_2 - \theta_1 \leq \dfrac{\pi}{2} - \theta_c$. We then have $x \cdot m = \cos(\theta_2 - \theta_1) \geq \cos(\pi/2 - \theta_c) = \kappa$, and therefore obtain the following physical constraint for refraction:

---

[2] If $\theta_1 > \theta_c$, then the phenomenon of total internal reflection occurs, see Fig. 1c.

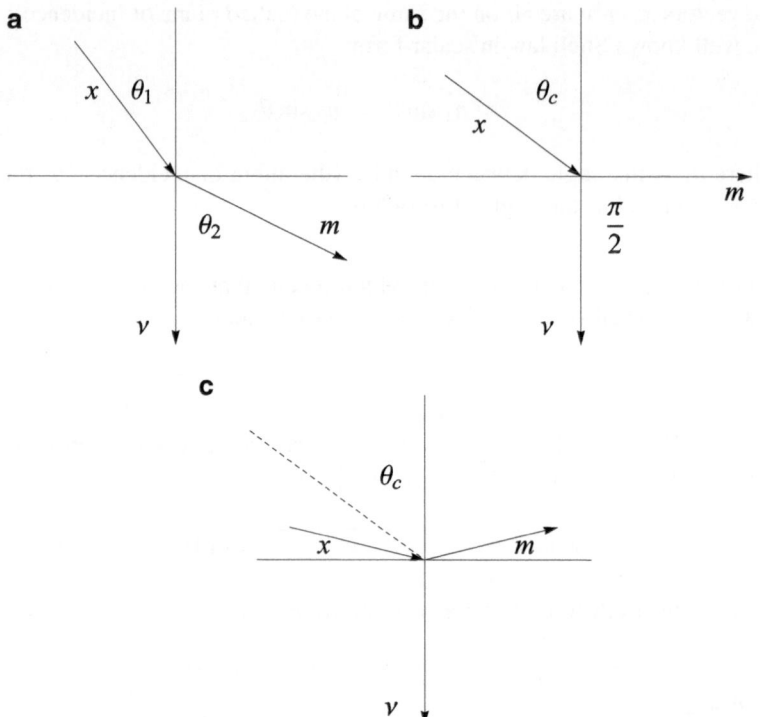

Fig. 1 Snell's law $\kappa < 1$, e.g., glass to air. (**a**) $x - \kappa m$ parallel to v. (**b**) Critical angle. (**c**) Internal reflection

$$\text{if } \kappa = n_2/n_1 < 1 \text{ and a ray of direction } x \text{ through medium } I$$

$$\text{is refracted into medium } II \text{ in the direction } m, \text{ then } m \cdot x \geq \kappa. \tag{5}$$

Notice also that in this case $\lambda > 0$ in (3).

Conversely, given $x, m \in S^2$ with $x \cdot m \geq \kappa$ and $\kappa < 1$, it follows from (4) that there exists a hyperplane refracting any ray through medium I with direction x into a ray of direction m in medium II.

2.1.2 $\kappa > 1$

In this case, waves propagate in medium I faster than in medium II, and the refracted rays tend to bent towards the normal. By the Snell law, the maximum angle of refraction denoted by θ_c^* is achieved when $\theta_1 = \pi/2$, and $\theta_c^* = \arcsin(1/\kappa)$. Once again from the Snell law in scalar form

$$\theta_1 - \theta_2 = \arcsin(\kappa \sin \theta_2) - \theta_2 \tag{6}$$

which is strictly increasing for $\theta_2 \in [0, \theta_c^*]$, and $0 \leq \theta_1 - \theta_2 \leq \dfrac{\pi}{2} - \theta_c^*$. We therefore obtain the following physical constraint for the case $\kappa > 1$:

if a ray with direction x traveling through medium I

is refracted into a ray in medium II with direction m, then $m \cdot x \geq 1/\kappa$. (7)

Notice also that in this case $\lambda < 0$ in (3).

On the other hand, by (6), if $x, m \in S^2$ with $x \cdot m \geq 1/\kappa$ and $\kappa > 1$, then there exists a hyperplane refracting any ray of direction x through medium I into a ray with direction m in medium II.

We summarize the above discussion on the physical constraints of refraction in the following lemma.

Lemma 2.1. *Let n_1 and n_2 be the indices of refraction of two media I and II, respectively, and $\kappa = n_2/n_1$. Then a light ray in medium I with direction $x \in S^2$ is refracted by some surface into a light ray with direction $m \in S^2$ in medium II if and only if $m \cdot x \geq \kappa$, when $\kappa < 1$; and if and only if $m \cdot x \geq 1/\kappa$, when $\kappa > 1$.*

2.1.3 $\kappa = 1$

This corresponds to reflection. It means

$$x - m = \lambda \, v. \tag{8}$$

Taking dot products with x and then with m yields $1 - m \cdot x = \lambda \, x \cdot v$ and $x \cdot m - 1 = \lambda \, m \cdot v$, then $x \cdot v = -m \cdot v$. Also taking dot product with v in (8) then yields $\lambda = 2 \, x \cdot v$. Therefore

$$m = x - 2(x \cdot v)v.$$

2.2 Derivation of the Snell Law

At time t, $\psi(x, y, z, t) = 0$ denotes a surface that separates the part of the space that is at rest with the part of the space that is disturbed by the electric and magnetic fields. For each t fixed the surface defined by $\psi(x, y, z, t) = 0$ is called a *wave front*. The *light rays* are the orthogonal trajectories to the wave fronts. We assume that $\psi_t \neq 0$ and so we can solve $\psi(x, y, z, t) = 0$ in t obtaining that

$$\phi(x, y, z) = ct,$$

where c is the speed of light in vacuum. Therefore, it t runs, then we get that the wave fronts are the level sets of the function $\phi(x, y, z)$.

Let us assume that the wave fronts travel in an homogenous and isotropic medium I with refractive index $n_1(= c/v_1)$, v_1 is the speed of propagation in medium I. This wave front is transmitted to another homogeneous and isotropic medium II having refractive index n_2. Let Σ be the surface in 3-d separating the media I and II, and suppose it is given by the equations $x = f(\xi, \eta)$, $y = g(\xi, \eta)$ and $z = h(\xi, \eta)$. Let $\phi_1(x, y, z) = ct$ be the wave front in medium I and $\phi_2(x, y, z) = ct$ be the wave front, to be determined, in medium II. On the surface Σ the two wave fronts agree, that is,

$$\phi_1\left(f(\xi, \eta), g(\xi, \eta), h(\xi, \eta)\right) = \phi_2\left(f(\xi, \eta), g(\xi, \eta), h(\xi, \eta)\right).$$

Differentiating this equation with respect to ξ and η yields

$$\left(\frac{\partial\phi_1}{\partial x} - \frac{\partial\phi_2}{\partial x}\right) f_\xi + \left(\frac{\partial\phi_1}{\partial y} - \frac{\partial\phi_2}{\partial y}\right) g_\xi + \left(\frac{\partial\phi_1}{\partial z} - \frac{\partial\phi_2}{\partial z}\right) h_\xi = 0$$

$$\left(\frac{\partial\phi_1}{\partial x} - \frac{\partial\phi_2}{\partial x}\right) f_\eta + \left(\frac{\partial\phi_1}{\partial y} - \frac{\partial\phi_2}{\partial y}\right) g_\eta + \left(\frac{\partial\phi_1}{\partial z} - \frac{\partial\phi_2}{\partial z}\right) h_\eta = 0.$$

This means that the vector $D\phi_1 - D\phi_2$ is perpendicular to both vectors (f_ξ, g_ξ, h_ξ) and (f_η, g_η, h_η), and therefore it is normal to the surface Σ. Let ν be the outer normal at the surface Σ. Then we have

$$(D\phi_1 - D\phi_2) = \lambda\, \nu, \tag{9}$$

for some scalar λ. A light ray $\gamma_1(t)$ in medium I has constant speed v_1 and a light ray $\gamma_2(t)$ in II constant speed v_2. Say $\gamma_1(t)$ is the light ray in medium I and $\gamma_2(t)$ the light ray in medium II. So we have $\phi_1(\gamma_1(t)) = ct$ and $\phi_2(\gamma_2(t)) = ct$. Differentiating with respect to t yields $D\phi_i(\gamma_i(t)) \cdot \gamma_i'(t) = c$, $i = 1, 2$. Let θ_i be the angle between the vectors $D\phi_i(\gamma_i(t)), \gamma_i'(t)$. Since the rays are orthogonal trajectories we get that $\theta_i = 0$ for $i = 1, 2$. If γ_i is parametrized so that $|\gamma_i'(t)| = v_i$, we then obtain that

$$|D\phi_i(\gamma_i(t))| = \frac{c}{v_i} = n_i.$$

If we let $x = \dfrac{D\phi_1(\gamma_1(t))}{|D\phi_1(\gamma_1(t))|}$ and $m = \dfrac{D\phi_2(\gamma_2(t))}{|D\phi_2(\gamma_2(t))|}$, we then obtain from (9) that

$$n_1\, x - n_2\, m = \lambda\, \nu$$

which is equivalent to (1).

2.3   Surfaces with the Uniform Refracting Property: Far Field Case

Let $m \in S^2$ be fixed, and we ask the following: if rays of light emanate from the origin inside medium I, what is the surface Γ, interface of the media I and II, that refracts all these rays into rays parallel to m?

Suppose Γ is parameterized by the polar representation $\rho(x)x$ where $\rho > 0$ and $x \in S^2$. Consider a curve on Γ given by $r(t) = \rho(x(t))x(t)$ for $x(t) \in S^2$. According to (2), the tangent vector $r'(t)$ to Γ satisfies $r'(t) \cdot (x(t) - \kappa m) = 0$. That is, $([\rho(x(t))]'x(t) + \rho(x(t))x'(t)) \cdot (x(t) - \kappa m) = 0$, which yields $(\rho(x(t))(1 - \kappa m \cdot x(t)))' = 0$. Therefore

$$\rho(x) = \frac{b}{1 - \kappa m \cdot x} \tag{10}$$

for $x \in S^2$ and for some $b \in \mathbb{R}$. To understand the surface given by (10), we distinguish two cases $\kappa < 1$ and $\kappa > 1$.

2.3.1   Case $\kappa < 1$

For $b > 0$, we will see that the surface Γ given by (10) is an ellipsoid of revolution about the axis of direction m. Suppose for simplicity that $m = e_n$, the nth-coordinate vector. If $y = (y', y_n) \in \mathbb{R}^n$ is a point on Γ, then $y = \rho(x)x$ with $x = y/|y|$. From (10), $|y| - \kappa y_n = b$, that is, $|y'|^2 + y_n^2 = (\kappa y_n + b)^2$ which yields $|y'|^2 + (1 - \kappa^2)y_n^2 - 2\kappa b y_n = b^2$. This surface Γ can be written in the form

$$\frac{|y'|^2}{\left(\dfrac{b}{\sqrt{1-\kappa^2}}\right)^2} + \frac{\left(y_n - \dfrac{\kappa b}{1-\kappa^2}\right)^2}{\left(\dfrac{b}{1-\kappa^2}\right)^2} = 1 \tag{11}$$

which is an ellipsoid of revolution about the y_n axis with foci $(0, 0)$ and $(0, 2\kappa b/(1 - \kappa^2))$. Since $|y| = \kappa y_n + b$ and the physical constraint for refraction (5), $\dfrac{y}{|y|} \cdot e_n \geq \kappa$ is equivalent to $y_n \geq \dfrac{\kappa b}{1 - \kappa^2}$. That is, for refraction to occur y must be in the upper part of the ellipsoid (11); we denote this semi-ellipsoid by $E(e_n, b)$, see Fig. 2. To verify that $E(e_n, b)$ has the uniform refracting property, that is, it refracts any ray emanating from the origin in the direction e_n, we check that (2) holds at each point. Indeed, if $y \in E(e_n, b)$, then $\left(\dfrac{y}{|y|} - \kappa e_n\right)\dfrac{y}{|y|} \geq 1 - \kappa > 0$, and $\left(\dfrac{y}{|y|} - \kappa e_n\right) \cdot e_n \geq 0$, and so $\dfrac{y}{|y|} - \kappa e_n$ is an outward normal to $E(e_n, b)$ at y.

Fig. 2 Only half of the
ellipsoid refracts in the
direction $m = e_n$

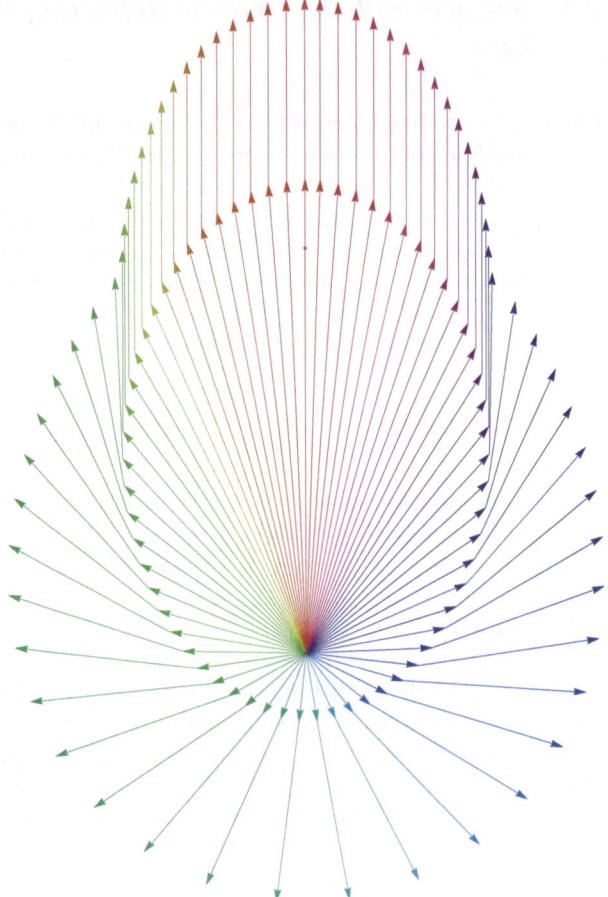

Rotating the coordinates, it is easy to see that the surface given by (10) with
$\kappa < 1$ and $b > 0$ is an ellipsoid of revolution about the axis of direction m with foci
0 and $\dfrac{2\kappa b}{1 - \kappa^2} m$. Moreover, the semi-ellipsoid $E(m, b)$ given by

$$E(m, b) = \{\rho(x)x : \rho(x) = \frac{b}{1 - \kappa\, m \cdot x}, \ x \in S^{n-1}, \ x \cdot m \geq \kappa\}, \quad (12)$$

has the uniform refracting property, any ray emanating from the origin O is refracted
in the direction m.

2.3.2 Case $\kappa > 1$

Due to the physical constraint of refraction (7), we must have $b < 0$ in (10). Define
for $b > 0$

$$H(m,b) = \{\rho(x)x : \rho(x) = \frac{b}{\kappa\, m \cdot x - 1}, \ x \in S^{n-1}, \ x \cdot m \geq 1/\kappa\}. \qquad (13)$$

We claim that $H(m,b)$ is the sheet with opening in direction m of a hyperboloid of revolution of two sheets about the axis of direction m. To prove the claim, set for simplicity $m = e_n$. If $y = (y', y_n) \in H(e_n, b)$, then $y = \rho(x)x$ with $x = y/|y|$. From (13), $\kappa\, y_n - |y| = b$, and therefore $|y'|^2 + y_n^2 = (\kappa\, y_n - b)^2$ which yields

$$|y'|^2 - (\kappa^2 - 1)\left[\left(y_n - \frac{\kappa b}{\kappa^2 - 1}\right)^2 - \left(\frac{\kappa b}{\kappa^2 - 1}\right)^2\right] = b^2. \ \text{Thus, any point } y \text{ on}$$

$H(e_n, b)$ satisfies the equation

$$\frac{\left(y_n - \dfrac{\kappa b}{\kappa^2 - 1}\right)^2}{\left(\dfrac{b}{\kappa^2 - 1}\right)^2} - \frac{|y'|^2}{\left(\dfrac{b}{\sqrt{\kappa^2 - 1}}\right)^2} = 1 \qquad (14)$$

which represents a hyperboloid of revolution of two sheets about the y_n axis with foci $(0,0)$ and $(0, 2\kappa b/(\kappa^2 - 1))$. Moreover, the upper sheet of this hyperboloid of revolution is given by

$$y_n = \frac{\kappa b}{\kappa^2 - 1} + \frac{b}{\kappa^2 - 1}\sqrt{1 + \frac{|y'|^2}{\left(b/\sqrt{\kappa^2 - 1}\right)^2}}$$

and satisfies $\kappa y_n - b > 0$, and hence has polar equation $\rho(x) = \dfrac{b}{\kappa\, e_n \cdot x - 1}$. Similarly, the lower sheet satisfies $\kappa y_n - b < 0$ and has polar equation $\rho(x) = \dfrac{b}{\kappa\, e_n \cdot x + 1}$. For a general m, by a rotation, we obtain that $H(m,b)$ is the sheet with opening in direction m of a hyperboloid of revolution of two sheets about the axis of direction m with foci $(0,0)$ and $\dfrac{2\kappa b}{\kappa^2 - 1}m$.

Notice that the focus $(0,0)$ is outside the region enclosed by $H(m,b)$ and the focus $\dfrac{2\kappa b}{\kappa^2 - 1}m$ is inside that region. The vector $\kappa m - \dfrac{y}{|y|}$ is an inward normal to $H(m,b)$ at y, because by (13)

$$\left(\kappa m - \frac{y}{|y|}\right) \cdot \left(\frac{2\kappa b}{\kappa^2 - 1}m - y\right) \geq \frac{2\kappa^2 b}{\kappa^2 - 1} - \frac{2\kappa b}{\kappa^2 - 1} - \kappa m \cdot y + |y|$$

$$= \frac{2\kappa b}{\kappa + 1} - b = \frac{b(\kappa - 1)}{\kappa + 1} > 0.$$

Clearly, $\left(\kappa m - \dfrac{y}{|y|}\right) \cdot m \geq \kappa - 1$ and $\left(\kappa m - \dfrac{y}{|y|}\right) \cdot \dfrac{y}{|y|} > 0$. Therefore, $H(m,b)$ satisfies the uniform refraction property.

We remark that one has to use $H(-e_n, b)$ to uniformly refract in the direction $-e_n$, and due to the physical constraint (7), the lower sheet of the hyperboloid of (14) cannot refract in the direction $-e_n$.

From the above discussion, we have proved the following.

Lemma 2.2. *Let n_1 and n_2 be the indexes of refraction of two media I and II, respectively, and $\kappa = n_2/n_1$. Assume that the origin O is inside medium I, and $E(m, b)$, $H(m, b)$ are defined by (12) and (13), respectively. We have:*

(i) *If $\kappa < 1$ and $E(m, b)$ is the interface of media I and II, then $E(m, b)$ refracts all rays emitted from O into rays in medium II with direction m.*

(ii) *If $\kappa > 1$ and $H(m, b)$ separates media I and II, then $H(m, b)$ refracts all rays emitted from O into rays in medium II with direction m.*

2.3.3   Case $\kappa = 1$

When $\kappa = 1$ we see this is a paraboloid. Indeed, let $m = -e_n$, then a point $X = \rho(x)x$ is on the surface (10) if $|X| = b - x_n$. The distance from X to the plane $x_n = b$ is $b - x_n$, and the distance from X to 0 is $|X|$. So this is a paraboloid with focus at 0, directrix plane $x_n = b$ and axis in the direction $-e_n$.

2.4   Uniform Refraction: Near Field Case

The question we ask is: given a point O inside medium I and a point P inside medium II, find an interface surface S between media I and II that refracts all rays emanating from the point O into the point P. Suppose O is the origin, and let $X(t)$ be a curve on S. By the Snell law of refraction the tangent vector $X'(t)$ satisfies

$$X'(t) \cdot \left(\frac{X(t)}{|X(t)|} - \kappa \frac{P - X(t)}{|P - X(t)|} \right) = 0.$$

That is,

$$|X(t)|' + \kappa |P - X(t)|' = 0.$$

Therefore S is the Cartesian oval

$$|X| + \kappa |X - P| = b. \tag{15}$$

Since $f(X) = |X| + \kappa |X - P|$ is a convex function, the oval is a convex set.

We need to find and analyze the polar equation of the oval. Write $X = \rho(x)x$ with $x \in S^{n-1}$. Then writing $\kappa |\rho(x)x - P| = b - \rho(x)$, squaring this quantity and solving the quadratic equation yields

$$\rho(x) = \frac{(b - \kappa^2 x \cdot P) \pm \sqrt{(b - \kappa^2 x \cdot P)^2 - (1 - \kappa^2)(b^2 - \kappa^2 |P|^2)}}{1 - \kappa^2}. \qquad (16)$$

Set

$$\Delta(t) = (b - \kappa^2 t)^2 - (1 - \kappa^2)(b^2 - \kappa^2 |P|^2). \qquad (17)$$

2.5 Case $0 < \kappa < 1$

We have

$$\Delta(x \cdot P) > \kappa^2 (x \cdot P - b)^2, \qquad \text{if } |x \cdot P| < |P|. \qquad (18)$$

If $b \geq |P|$, then O and P are inside or on the oval, and so the oval cannot refract rays to P. If the oval is non empty, then $\kappa |P| \leq b$. In case $\kappa |P| = b$, the oval reduces to the point O. The only interesting case is then $\kappa |P| < b < |P|$. From the equation of the oval we get that $\rho(x) \leq b$. So we now should decide which values $\pm$ to take in the definition of $\rho(x)$. Let ρ_+ and ρ_- be the corresponding ρ's. We claim that $\rho_+(x) > b$ and $\rho_-(x) \leq b$. Indeed,

$$
\begin{aligned}
\rho_+(x) &= \frac{(b - \kappa^2 x \cdot P) + \sqrt{\Delta(x \cdot P)}}{1 - \kappa^2} \\
&\geq \frac{(b - \kappa^2 x \cdot P) + \kappa |b - x \cdot P|}{1 - \kappa^2} \\
&= b + \frac{\kappa^2 (b - x \cdot P) + \kappa |b - x \cdot P|}{1 - \kappa^2} \\
&\geq b.
\end{aligned}
$$

The equality $\rho_+(x) = b$ holds only if $|x \cdot P| = |P|$ and $b = x \cdot P$. So $\rho_+(x) > b$ if $\kappa |P| < b < |P|$. Similarly,

$$
\begin{aligned}
\rho_-(x) &= \frac{(b - \kappa^2 x \cdot P) - \sqrt{\Delta(x \cdot P)}}{1 - \kappa^2} \\
&\leq \frac{(b - \kappa^2 x \cdot P) - \kappa |b - x \cdot P|}{1 - \kappa^2} \\
&= b + \frac{\kappa^2 (b - x \cdot P) - \kappa |b - x \cdot P|}{1 - \kappa^2} \\
&\leq b.
\end{aligned}
$$

So the claim is proved. Therefore the polar equation of the oval is then given by

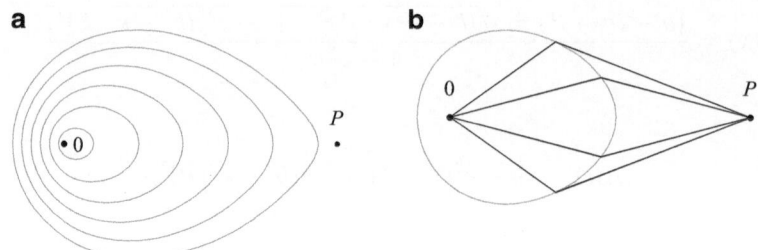

Fig. 3 Cartesian ovals $\kappa < 1$, e.g., glass to air. (**a**) $|X| + 2/3|X - P| = 1.4 - 1.9$, $P = (2,0)$. (**b**) $|X| + 2/3|X - P| = 1.7$, $P = (2,0)$

$$h(x, P, b) = \rho_-(x) = \frac{(b - \kappa^2 x \cdot P) - \sqrt{\Delta(x \cdot P)}}{1 - \kappa^2}. \qquad (19)$$

From the physical constraint for refraction we must have $x \cdot \left(\dfrac{P - h(x,P,b)x}{|P - h(x,P,b)x|} \right) \geq \kappa$, and from the equation of the oval we then get that to have refraction we need

$$x \cdot P \geq b. \qquad (20)$$

Concluding with the case $\kappa < 1$, given $P \in \mathbb{R}^n$ and $\kappa|P| < b < |P|$, keeping in mind (19) and (20), a refracting oval is the set (Fig. 3)

$$\mathcal{O}(P, b) = \{ h(x, P, b) \, x : x \in S^{n-1}, x \cdot P \geq b \}$$

where

$$h(x, P, b) = \frac{(b - \kappa^2 x \cdot P) - \sqrt{(b - \kappa^2 x \cdot P)^2 - (1 - \kappa^2)(b^2 - \kappa^2 |P|^2)}}{1 - \kappa^2}.$$

Remark 2.3. If $|P| \to \infty$, then the oval converges to an ellipsoid which is the surface having the uniform refraction property in the far field case, see Sect. 2.3.1. In fact, if $m = P/|P|$ and $b = \kappa|P| + C$ with C positive constant we have

$$h(x, P, b) = \frac{b^2 - \kappa^2 |P|^2}{b - \kappa^2 x \cdot P + \sqrt{\Delta(x \cdot P)}}$$

$$= \frac{C(2\kappa|P| + C)}{(\kappa|P| - \kappa^2 x \cdot m|P| + C) + \sqrt{(\kappa|P| - \kappa^2 x \cdot m|P| + C)^2 - (1 - \kappa^2)C(2\kappa|P| + C)}}$$

$$\to \frac{2\kappa C}{(\kappa - \kappa^2 x \cdot m) + \sqrt{(\kappa - \kappa^2 x \cdot m)^2}} = \frac{C}{1 - \kappa x \cdot m}$$

as $|P| \to \infty$.

2.6 Case $\kappa > 1$

In this case we must have $|P| \leq b$, and in case $b = |P|$ the oval reduces to the point P. Also $b < \kappa |P|$, since otherwise the points 0, P are inside the oval or 0 is on the oval, and therefore there cannot be refraction if $b \geq \kappa |P|$. So to have refraction we must have $|P| < b < \kappa |P|$ and so the point P is inside the oval and 0 is outside the oval.

Rewriting ρ in (16) we get that

$$\rho_{\pm}(x) = \frac{(\kappa^2 x \cdot P - b) \pm \sqrt{(\kappa^2 x \cdot P - b)^2 - (\kappa^2 - 1)(\kappa^2 |P|^2 - b^2)}}{\kappa^2 - 1}$$

for $\Delta(x \cdot P) \geq 0$ which amounts $x \cdot P \geq \dfrac{b + \sqrt{(\kappa^2 - 1)(\kappa^2 |P|^2 - b^2)}}{\kappa^2}$. Notice that $\rho_{\pm}(x) < 0$ for $\kappa^2 x \cdot P - b < 0$. We have that $\rho_-(x) \leq \rho_+(x) \leq \dfrac{(\kappa^2 |P| - b) + \sqrt{\Delta(|P|)}}{\kappa^2 - 1} = \dfrac{\kappa |P| + b}{\kappa + 1} < b$. To have refraction, by the physical constraint we need to have $x \cdot \dfrac{P - x \rho_{\pm}(x)}{|P - x \rho_{\pm}(x)|} \geq 1/\kappa$, which is equivalent to $\kappa^2 x \cdot P - b \geq (\kappa^2 - 1)\rho_{\pm}(x)$. Therefore, the physical constraint is satisfied only by ρ_-.

Therefore if $\kappa > 1$, refraction only occurs when $|P| < b < \kappa |P|$, and the refracting piece of the oval is then given by

$$\mathcal{O}(P, b) = \left\{ h(x, P, b)x : x \cdot P \geq \frac{b + \sqrt{(\kappa^2 - 1)(\kappa^2 |P|^2 - b^2)}}{\kappa^2} \right\} \qquad (21)$$

with

$$h(x, P, b) = \rho_-(x) = \frac{(\kappa^2 x \cdot P - b) - \sqrt{(\kappa^2 x \cdot P - b)^2 - (\kappa^2 - 1)(\kappa^2 |P|^2 - b^2)}}{\kappa^2 - 1},$$
$$(22)$$

see (Fig. 4).

Remark 2.4. If $|P| \to \infty$, then the oval $\mathcal{O}(P, b)$ converges to the semi hyperboloid appearing in the far field refraction problem when $\kappa > 1$, see Sect. 2.3.2. Indeed, let $m = \dfrac{P}{|P|} \in S^{n-1}$ and $b = \kappa |P| - a$ with $a > 0$ a constant. For $x \in \Gamma(P, b)$ we have

$$\frac{b + \sqrt{(\kappa^2 - 1)(\kappa^2 |P|^2 - b^2)}}{\kappa^2 |P|} = \frac{\kappa |P| - a + \sqrt{(\kappa^2 - 1)(\kappa^2 |P|^2 - (\kappa |P| - a)^2)}}{\kappa^2 |P|} \to \frac{1}{\kappa}$$

as $|P| \to \infty$. On the other hand, if $x \cdot m > 1/\kappa$, we get

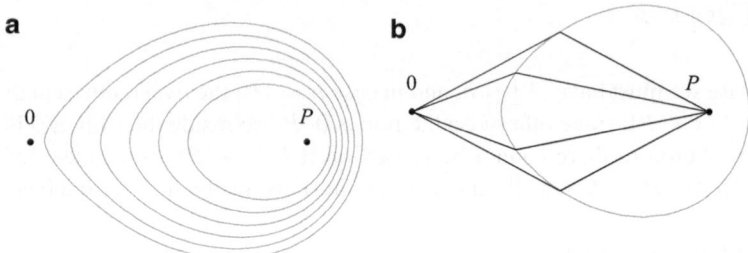

Fig. 4 Cartesian ovals $\kappa > 1$, e.g., air to glass. (**a**) $|X| + 3/2|X - P| = 2.9 - 2.4$, $P = (2, 0)$. (**b**) $|X| + 3/2|X - P| = 2.7$, $P = (2, 0)$

$$h(x, P, b) = \frac{a(2\kappa|P| - a)}{(\kappa^2|P|x \cdot m - \kappa|P| + a) + \sqrt{(\kappa^2|P|x \cdot m - \kappa|P| + a)^2 - (\kappa^2 - 1)a(2\kappa|P| - a)}}$$

$$\rightarrow \frac{a2\kappa}{\kappa^2 x \cdot m - \kappa + \sqrt{(\kappa^2 x \cdot m - \kappa)^2}} = \frac{a}{\kappa x \cdot m - 1},$$

as $|P| \rightarrow \infty$.

3   Optimal Mass Transportation

Let D, D^* be two domains on S^{n-1} or domains in a manifold (D might be contained in one manifold and D^* in another) with $|\partial D| = 0$.

Let $\mathcal{N}$ be a multi-valued mapping from $\overline{D}$ onto $\overline{D^*}$ such that $\mathcal{N}(x)$ is single-valued a.e. on $\overline{D}$. We also assume that $|\{x \in \overline{D} : \mathcal{N}(x) = \emptyset\}| = 0$. For $F \subset \overline{D^*}$, we set

$$\mathcal{T}(F) = \mathcal{N}^{-1}(F) = \{x \in \overline{D} : \mathcal{N}(x) \cap F \neq \emptyset\}.$$

We say $\mathcal{N}$ is measurable if $\mathcal{T}(F)$ is Lebesgue measurable for any Borel set $F \subset \overline{D^*}$.

For example, if $\mathcal{N} = \partial u$ with u convex, then $\mathcal{N}$ is measurable. Because for u convex we have $m \in \partial u(x_0)$ iff $x_0 \in \partial u^*(m)$, where u^* is the Legendre transformation of u. Therefore $(\partial u)^{-1}(F) = \partial u^*(F)$, and since $\partial u^*(F)$ is Lebesgue measurable for each Borel set F, we get that ∂u is measurable.

Suppose $g \in L^1(D)$ is nonnegative and Γ on $\overline{D^*}$ is a finite Radon measure satisfying the conservation condition

$$\int_{\overline{D}} g(x)\, dx = \Gamma(\overline{D^*}) > 0. \tag{1}$$

Notice that the set function $\mu(F) = \int_{\mathcal{T}(F)} g(x)\, dx$ is Borel measure because $\mathcal{N}$ is single valued a.e., and measurable. Indeed, if S is the set of measure zero such that $\mathcal{N}(x)$ is single valued for all $x \in \overline{D} \setminus S$, and F_j is a sequence of disjoint Borel sets, then $|T(F_i) \cap T(F_j)| = 0$ for $i \neq j$, then μ is σ-additive. Therefore μ is a finite Borel measure and so is regular.

We say $\mathcal{N}$ is measure preserving from $g(x)dx$ to Γ if for any Borel $F \subset \overline{D^*}$

$$\int_{T(F)} g(x)\, dx = \Gamma(F). \tag{2}$$

Lemma 3.1. $\mathcal{N}$ is a measure preserving mapping from $g(x)dx$ to Γ if and only if for any $v \in C(\overline{D^*})$

$$\int_{\overline{D}} v(\mathcal{N}(x))g(x)\, dx = \int_{\overline{D^*}} v(m)\, d\,\Gamma(m). \tag{3}$$

We remark that $v(\mathcal{N}(x))$ is well defined for $x \in \overline{D} \setminus S$ where $\mathcal{N}(x)$ is single-valued on $\overline{D} \setminus S$ and $|S| = 0$, and $\int_{\overline{D}} v(\mathcal{N}(x))g(x)\, dx$ is understood as $\int_{\overline{D} \setminus S} v(\mathcal{N}(x))g(x)\, dx$.

Proof. Let $\mathcal{N}$ be a measure preserving mapping. To show (3), it suffices to prove it for $v = \chi_F$, the characteristic function of a Borel set F, because for each v continuous there exists a sequence of simple functions converging uniformly to v. It is easy to verify that $\chi_{T(F)}(x) = \chi_F(\mathcal{N}(x))$ for $x \in \overline{D} \setminus S$. Therefore by (2)

$$\int_{\overline{D^*}} \chi_F(m)\, d\Gamma = \int_{T(F) \cap (\overline{D} \setminus S)} g\, dx = \int_{\overline{D} \setminus S} \chi_F(\mathcal{N}(x))g(x)\, dx.$$

To prove the converse, assume that (3) holds. We will show that for any Borel set $E \subset \overline{D^*}$

$$\int_{T(E)} g\, dx \leq \Gamma(E). \tag{4}$$

Indeed, let us first assume that $E = G$ is open, then given a compact set $K \subset G$, choose $v \in C(\overline{D^*})$ such that $0 \leq v \leq 1$, $v = 1$ on K, and $v = 0$ outside G. By (3), one gets

$$\int_{T(K)} g(x)\, dx \leq \int_{\overline{D}} v(\mathcal{N}(x))g(x)\, dx \leq \Gamma(G),$$

for each compact $K \subset G$. Since μ is regular, (4) follows for E open. For a general Borel set $E \subset \overline{D^*}$, since Γ is also regular, given $\epsilon > 0$ there exists G open $E \subset G$ with $\Gamma(G \setminus E) < \epsilon$. Then

$$\int_{T(E)} g(x)\, dx \leq \int_{T(G)} g(x)\, dx \leq \Gamma(G) = \Gamma(E) + \Gamma(G \setminus E) < \Gamma(E) + \epsilon$$

and so (4) follows.

We next prove that equality holds in (4).
First notice that

$$\{x \in \bar{D} : \mathcal{N}(x) \neq \emptyset\} \cap (\mathcal{T}(F))^c \subset \mathcal{T}(F^c),$$

for any set $F \subset \bar{D}^*$. Then applying (4) to $\overline{D^*} \setminus F$ with F Borel set yields

$$\int_{\{x \in \bar{D}:\mathcal{N}(x)\neq\emptyset\}\cap(\mathcal{T}(F))^c} g(x)\,dx \leq \int_{\mathcal{T}(F^c)} g(x)\,dx \leq \Gamma(\overline{D^*} \setminus F) = \Gamma(\overline{D^*}) - \Gamma(F).$$

Since $|\{x \in \bar{D} : \mathcal{N}(x) = \emptyset\}| = 0$, we have

$$\int_{\{x \in \bar{D}:\mathcal{N}(x)\neq\emptyset\}\cap(\mathcal{T}(F))^c} g(x)\,dx = \int_{(\mathcal{T}(F))^c} g(x)\,dx = \int_{\bar{D}} g(x)\,dx - \int_{\mathcal{T}(F)} g(x)\,dx.$$

So from the conservation condition (1), we obtain the reverse inequality in (4). $\square$

Consider the general cost function $c(x, m) \in Lip(\overline{D} \times \overline{D^*})$, the space of Lipschitz functions on $\overline{D} \times \overline{D^*}$, and *the set of admissible functions*

$$\mathcal{K} = \{(u, v) : u \in C(\overline{D}), v \in C(\overline{D^*}), u(x) + v(m) \leq c(x, m), \forall x \in D, \forall m \in D^*\}.$$

Define the *dual functional* I for $(u, v) \in C(\overline{D}) \times C(\overline{D^*})$

$$I(u, v) = \int_D u(x) g(x)\,dx + \int_{\overline{D^*}} v(m)\,d\Gamma,$$

and define the c- *and* c^*-*transforms*

$$u^c(m) = \inf_{x \in \overline{D}} [c(x, m) - u(x)], \quad m \in \overline{D^*}; \qquad v_c(x) = \inf_{m \in \overline{D^*}} [c(x, m) - v(m)], \quad x \in \overline{D}.$$

Definition 3.2. A function $\phi \in C(\overline{D})$ is c-concave if for $x_0 \in \overline{D}$, there exist $m_0 \in \overline{D^*}$ and $b \in \mathbb{R}$ such that $\phi(x) \leq c(x, m_0) - b$ on $\overline{D}$ with equality at $x = x_0$.

Obviously v_c is c-concave for any $v \in C(\overline{D^*})$. We collect the following properties:

1. For any $u \in C(\overline{D})$ and $v \in C(\overline{D^*})$, $v_c \in Lip(\overline{D})$ and $u^c \in Lip(\overline{D^*})$ with Lipschitz constants bounded uniformly by the Lipschitz constant of c. Indeed, (x_0 the point where the minimum is attained)

$$u^c(m_1) - u^c(m_2) \leq u^c(m_1) - (c(x_0, m_2) - u(x_0))$$

$$\leq c(x_0, m_1) - u(x_0) - c(x_0, m_2) + u(x_0) \leq K|m_1 - m_2|.$$

2. If $(u, v) \in \mathcal{K}$, then $v(m) \leq u^c(m)$ and $u(x) \leq v_c(x)$. Also $(v_c, v), (u, u^c) \in \mathcal{K}$.
3. ϕ is c-concave iff $\phi = (\phi^c)_c$.

   Indeed, if $\phi(x) \leq c(x, m_0) - b$ on $\overline{D}$ and the equality holds at $x = x_0$, then $b = \phi^c(m_0)$. So $\phi(x_0) = c(x_0, m_0) - \phi^c(m_0)$ which yields $\phi(x_0) \geq (\phi^c)_c(x_0)$. On the other hand, from the definitions of c and c^* transforms we always have that $(\phi^c)_c \geq \phi$ for any ϕ.

Definition 3.3. Given a function $\phi(x)$, the c-normal mapping of ϕ is defined by

$$\mathcal{N}_{c,\phi}(x) = \{m \in \overline{D^*} : \phi(x) + \phi^c(m) = c(x, m)\}, \qquad \text{for } x \in \overline{D},$$

and $\mathcal{T}_{c,\phi}(m) = \mathcal{N}_{c,\phi}^{-1}(m) = \{x \in \overline{D} : m \in \mathcal{N}_{c,\phi}(x)\}$.

We assume that the cost function $c(x, m)$ satisfies the following:

For any c-concave function ϕ, $\mathcal{N}_{c,\phi}(x)$ is single-valued a.e. on $\overline{D}$ (5)

and $\mathcal{N}_{c,\phi}$ is Lebesgue measurable.

Notice that if $c(x, m) = x \cdot m$, then $\mathcal{N}_{c,\phi}(x) = \partial^*\phi(x)$, where $\partial^*\phi$ is the super-differential of ϕ

$$\partial^*\phi(x) = \{m \in \mathbb{R}^n : \phi(y) \leq \phi(x) + m \cdot (y - x) \,\forall\, y \in \Omega\},$$

and we have $\partial^*\phi(x) = -\partial(-\phi)(x)$.

Lemma 3.4. *Suppose that $c(x, m)$ satisfies the assumption (5). Then*

(i) *If ϕ is c-concave and $\mathcal{N}_{c,\phi}$ is measure preserving from $g(x)dx$ to Γ, then (ϕ, ϕ^c) is a maximizer of $I(u, v)$ in $\mathcal{K}$.*
(ii) *If $\phi(x)$ is c-concave and (ϕ, ϕ^c) maximizes $I(u, v)$ in $\mathcal{K}$, then $\mathcal{N}_{c,\phi}$ is measure preserving from $g(x)dx$ to Γ.*

Proof. First prove (i). Given $(u, v) \in \mathcal{K}$, obviously

$$u(x) + v(\mathcal{N}_{c,\phi}(x)) \leq c(x, \mathcal{N}_{c,\phi}(x)) = \phi(x) + \phi^c(\mathcal{N}_{c,\phi}(x)), \qquad \text{a.e. } x \text{ on } \overline{D}.$$

Integrating the above inequality with respect to $g\,dx$ yields

$$\int_{\overline{D}} ug\,dx + \int_{\overline{D}} v(\mathcal{N}_{c,\phi}(x))g(x)\,dx \leq \int_{\overline{D}} \phi g\,dx + \int_{\overline{D}} \phi^c(\mathcal{N}_{c,\phi}(x))g(x)\,dx.$$

By Lemma 3.1, it yields $I(u, v) \leq I(\phi, \phi^c)$ and from (2) above the conclusion follows.

To prove (ii), let $\psi = \phi^c$, and for $v \in C(\overline{D^*})$, let $\psi_\theta(m) = \psi(m) + \theta\,v(m)$ where $-\epsilon_0 < \theta \leq \epsilon_0$ with ϵ_0 small, and let $\phi_\theta = (\psi_\theta)_c$. We shall prove that

$$\lim_{\theta \to 0} \frac{I(\phi_\theta, \psi_\theta) - I(\phi, \psi)}{\theta} = \int_{\overline{D}} -v(\mathcal{N}_{c,\phi}(x))\,g\,dx + \int_{\overline{D^*}} v(m)\,d\Gamma. \qquad (6)$$

Since $(\phi_\theta, \psi_\theta) \in \mathcal{K}$, we have $I(\phi_\theta, \psi_\theta) \leq I(\phi, \psi)$ for $-\epsilon_0 < \theta \leq \epsilon_0$, and hence the existence of the limit (6) implies it must be zero. Therefore the measure preserving property of $N_{c,\phi}$ follows from Lemma 3.1.

To prove (6) we write

$$\frac{I(\phi_\theta, \psi_\theta) - I(\phi, \psi)}{\theta} = \int_{\overline{D}} \frac{\phi_\theta - \phi}{\theta} g\, dx + \int_{\overline{D^*}} v(m)\, d\Gamma.$$

By Lebesgue dominated convergence theorem, to show (6), it is enough to show that $\dfrac{\phi_\theta(x) - \phi(x)}{\theta}$ is uniformly bounded, and $\dfrac{\phi_\theta(x) - \phi(x)}{\theta} \to -v(N_{c,\phi}(x))$ for all $x \in D \setminus S$, where $N_{c,\phi}(x)$ is single-valued on $D \setminus S$ and $|S| = 0$. Let us first prove the uniform boundedness. Fix $x \in \overline{D}$, we have by continuity that $\phi_\theta(x) = c(x, m_\theta) - \psi_\theta(m_\theta)$ for some $m_\theta \in \overline{D^*}$. Since ϕ is c-concave there exists $m_1 \in \overline{D^*}$ and $b \in \mathbb{R}$ such that $\phi(y) \leq c(y, m_1) - b$ for all $y \in \overline{D}$ with equality when $y = x$. This implies that $b = \phi^c(m_1)$ and so $\phi(x) = c(x, m_1) - \psi(m_1)$. Hence

$$\phi_\theta(x) - \phi(x) = c(x, m_\theta) - \psi(m_\theta) - \theta\, v(m_\theta) - \phi(x)$$
$$\geq \psi_c(x) - \theta\, v(m_\theta) - \phi(x) = (\phi^c)_c(x) - \theta\, v(m_\theta) - \phi(x) \geq -\theta\, v(m_\theta),$$

by (3) above. We also have

$$\phi_\theta(x) - \phi(x) = \phi_\theta(x) - c(x, m_1) + \psi(m_1) = \phi_\theta(x) - c(x, m_1) + \psi_\theta(m_1) - \theta\, v(m_1)$$
$$\leq \phi_\theta(x) - (\psi_\theta)_c(x) - \theta\, v(m_1) = -\theta\, v(m_1).$$

Then we get

$$-\theta\, v(m_\theta) \leq \phi_\theta(x) - \phi(x) \leq -\theta\, v(m_1).$$

Moreover, if $x \in \overline{D} \setminus S$, then $m_1 = N_{c,\phi}(x)$ since $\psi = \phi^c$. To finish the proof, we show that m_θ converges to m_1 as $\theta \to 0$. Otherwise, there exists a sequence m_{θ_k} such that $m_{\theta_k} \to m_\infty \neq m_1$. So $\phi(x) = \lim_{\theta \to 0} \phi_\theta(x) = c(x, m_\infty) - \psi(m_\infty)$, which yields $m_\infty \in N_{c,\phi}(x)$. We then get $m_1 = m_\infty$, a contradiction. The proof is complete. $\qquad\square$

Lemma 3.5. *There exists a c-concave ϕ such that*

$$I(\phi, \phi^c) = \sup\{I(u, v) : (u, v) \in \mathcal{K}\}.$$

Proof. Let

$$I_0 = \sup\{I(u, v) : (u, v) \in \mathcal{K}\},$$

and let $(u_k, v_k) \in \mathcal{K}$ be a sequence such that $I(u_k, v_k) \to I_0$. Set $\bar{u}_k = (v_k)_c$ and $\bar{v}_k = (\bar{u}_k)^c$. From property (2) above, $(\bar{u}_k, \bar{v}_k) \in \mathcal{K}$, $u_k \leq \bar{u}_k$, $v_k \leq \bar{v}_k$, and so $I(\bar{u}_k, \bar{v}_k) \to I_0$. Let $c_k = \min_{\overline{D}} \bar{u}_k$ and define

$$u_k^{\sharp} = \bar{u}_k - c_k, \qquad v_k^{\sharp} = \bar{v}_k + c_k.$$

Obviously $(u_k^{\sharp}, v_k^{\sharp}) \in \mathcal{K}$ and by the mass conservation condition on $g\,dx$ and Γ, equation (2), $I(\bar{u}_k, \bar{v}_k) = I(u_k^{\sharp}, v_k^{\sharp})$. Since $\bar{u}_k$ are uniformly Lipschitz, $u_k^{\sharp}$ are uniformly bounded. In addition, $v_k^{\sharp} = (\bar{u}_k)^c + c_k = (u_k^{\sharp})^c$ and consequently $v_k^{\sharp}$ are also uniformly bounded. By Arzelá-Ascoli's theorem, $(u_k^{\sharp}, v_k^{\sharp})$ contains a subsequence converging uniformly to (ϕ, ψ) on $\overline{D} \times \overline{D}^*$. We then obtain that $(\phi, \psi) \in \mathcal{K}$ and $I_0 = \sup\{I(u, v) : (u, v) \in \mathcal{K}\} = I(\phi, \psi)$. Notice that this shows in particular that the supremum of I over $\mathcal{K}$ is finite. From property (2) above, $(\psi_c, (\psi_c)^c)$ is the sought maximizer of $I(u, v)$, and ψ_c is c-concave.   □

Lemma 3.6. *Suppose that $c(x, m)$ satisfies the assumption (5). Let (ϕ, ϕ^c) with ϕ c-concave be a maximizer of $I(u, v)$ in $\mathcal{K}$. Then $\displaystyle\inf_{s \in S} \int_{\overline{D}} c(x, s(x))g(x)\,dx$ is attained at $s = \mathcal{N}_{c,\phi}$, where S is the class of measure preserving mappings from $g(x)dx$ to Γ. Moreover*

$$\inf_{s \in S} \int_{\overline{D}} c(x, s(x))g(x)\,dx = \sup\{I(u, v) : (u, v) \in \mathcal{K}\}. \qquad (7)$$

Proof. Let $\psi = \phi^c$. For $s \in S$, we have

$$\int_{\overline{D}} c(x, s(x))g(x)\,dx \geq \int_{\overline{D}} (\phi(x) + \psi(s(x)))\,g(x)\,dx$$

$$= \int_{\overline{D}} \phi(x)g(x)\,dx + \int_{\overline{D}} \psi(s(x))g(x)\,dx$$

$$= \int_{\overline{D}} \phi(x)g(x)\,dx + \int_{\overline{D}^*} \psi(m)\,d\Gamma = I(\phi, \psi)$$

$$= \int_{\overline{D}} (\phi(x) + \psi(\mathcal{N}_{c,\phi}(x)))\,g(x)\,dx, \quad \text{from Lemma 3.4(ii)}$$

$$= \int_{\overline{D}} c(x, \mathcal{N}_{c,\phi}(x))g(x)\,dx.$$

□

Obviously, for any c-concave function ϕ, $\mathcal{N}_{c,\phi}$ has the following converging property (C): if $m_k \in \mathcal{N}_{c,\phi}(x_k)$, $x_k \longrightarrow x_0$ and $m_k \longrightarrow m_0$, then $m_0 \in \mathcal{N}_{c,\phi}(x_0)$.

Lemma 3.7. *Assume that $c(x, m)$ satisfies the assumption (5) and that $\int_G g\,dx > 0$ for any open $G \subset D$. Then the minimizing mapping of $\displaystyle\inf_{s \in S} \int_{\overline{D}} c(x, s(x))g(x)\,dx$ is unique in the class of measure preserving mappings from $g(x)dx$ to Γ with the converging property (C).*

Proof. From Lemmas 3.5 and 3.6, let $\mathcal{N}_{c,\phi}$ be a minimizing mapping associated with a maximizer (ϕ, ϕ^c) of $I(u, v)$ with ϕ c-concave. Suppose that $\mathcal{N}_0$ is another minimizing mapping with the converging property (C). Clearly

$$\int_{\overline{D}} \left(c(x, \mathcal{N}_0(x)) - \phi(x) - \phi^c(\mathcal{N}_0(x)) \right) g(x) \, dx$$

$$= \inf_{s \in \mathcal{S}} \int_{\overline{D}} c(x, s(x)) g(x) \, dx - \left(\int_{\overline{D}} \phi(x) g(x) \, dx + \int_{\overline{D^*}} \phi^c(m) \, d\Gamma \right) = 0,$$

and since $\phi(x) + \phi^c(\mathcal{N}_0(x)) \leq c(x, \mathcal{N}_0(x))$, it follows that $\phi(x) + \phi^c(\mathcal{N}_0(x)) = c(x, \mathcal{N}_0(x))$ on the set $\{x \in D : g(x) > 0\}$ which is dense in D. Hence from (5) and the converging property (C), we get $\mathcal{N}_0(x) = \mathcal{N}_{c,\phi}(x)$ a.e. on D. $\square$

We remark from the above proof that if $g(x) > 0$ on D, then the minimizing mapping of $\inf_{s \in \mathcal{S}} \int_{\overline{D}} c(x, s(x)) g(x) \, dx$ is unique in the class of measure preserving mappings from $g(x) dx$ to Γ.

3.1 Application to the Refractor Problem $\kappa < 1$

Let n_1 and n_2 be the indexes of refraction of two homogeneous and isotropic media I and II, respectively. Suppose that from a point O inside medium I light emanates with intensity $f(x)$ for $x \in \Omega$. We want to construct a refracting surface $\mathcal{R}$ parameterized as $\mathcal{R} = \{\rho(x)x : x \in \overline{\Omega}\}$, separating media I and II, and such that all rays refracted by $\mathcal{R}$ into medium II have directions in Ω^* and the prescribed illumination intensity received in the direction $m \in \Omega^*$ is $f^*(m)$.

We first introduce the notions of refractor mapping and measure, and weak solution. In the next section we then convert the refractor problem into an optimal mass transport problem from $\overline{\Omega}$ to $\overline{\Omega^*}$ with the cost function $\log \dfrac{1}{1 - \kappa\, x \cdot m}$ and establish existence and uniqueness of weak solutions.

Let Ω, Ω^* be two domains on S^{n-1}, the illumination intensity of the emitting beam is given by nonnegative $f(x) \in L^1(\overline{\Omega})$, and the prescribed illumination intensity of the refracted beam is given by a nonnegative Radon measure μ on $\overline{\Omega^*}$. Throughout this section, we assume that $|\partial\Omega| = 0$ and the physical constraint

$$\inf_{x \in \Omega, m \in \Omega^*} x \cdot m \geq \kappa. \tag{1}$$

We further suppose that the total energy conservation

$$\int_{\Omega} f(x) \, dx = \mu(\overline{\Omega^*}) > 0, \tag{2}$$

and for any open set $G \subset \Omega$

$$\int_G f(x)\, dx > 0, \tag{3}$$

where dx denotes the surface measure on S^{n-1}.

3.2 Refractor Measure and Weak Solutions

We begin with the notions of refractor and supporting semi-ellipsoid.

Definition 3.8. A surface $\mathcal{R}$ parameterized by $\rho(x)x$ with $\rho \in C(\overline{\Omega})$ is a refractor from $\overline{\Omega}$ to $\overline{\Omega^*}$ for the case $\kappa < 1$ (often simply called as refractor in this section) if for any $x_0 \in \overline{\Omega}$ there exists a semi-ellipsoid $E(m,b)$ with $m \in \overline{\Omega^*}$ such that
$$\rho(x_0) = \frac{b}{1 - \kappa\, m \cdot x_0} \quad \text{and} \quad \rho(x) \le \frac{b}{1 - \kappa\, m \cdot x} \quad \text{for all } x \in \overline{\Omega}. \text{ Such } E(m,b) \text{ is}$$
called a supporting semi-ellipsoid of $\mathcal{R}$ at the point $\rho(x_0)x_0$.

From the definition, any refractor is globally Lipschitz on $\overline{\Omega}$.

Definition 3.9. Given a refractor $\mathcal{R} = \{\rho(x)x : x \in \overline{\Omega}\}$, the refractor mapping of $\mathcal{R}$ is the multi-valued map defined by for $x_0 \in \overline{\Omega}$

$$\mathcal{N}_{\mathcal{R}}(x_0) = \{m \in \overline{\Omega^*} : E(m,b) \text{ supports } \mathcal{R} \text{ at } \rho(x_0)x_0 \text{ for some } b > 0\}.$$

Given $m_0 \in \overline{\Omega^*}$, the tracing mapping of $\mathcal{R}$ is defined by

$$\mathcal{T}_{\mathcal{R}}(m_0) = \mathcal{N}_{\mathcal{R}}^{-1}(m_0) = \{x \in \overline{\Omega} : m_0 \in \mathcal{N}_{\mathcal{R}}(x)\}.$$

Definition 3.10. Given a refractor $\mathcal{R} = \{\rho(x)x : x \in \overline{\Omega}\}$, the Legendre transform of $\mathcal{R}$ is defined by

$$\mathcal{R}^* = \{\rho^*(m)m : \rho^*(m) = \inf_{x \in \overline{\Omega}} \frac{1}{\rho(x)(1 - \kappa\, x \cdot m)}, \ m \in \overline{\Omega^*}\}.$$

We now give some basic properties of Legendre transforms.

Lemma 3.11. *Let $\mathcal{R}$ be a refractor from $\overline{\Omega}$ to $\overline{\Omega^*}$. Then*

(i) $\mathcal{R}^*$ is a refractor from $\overline{\Omega^*}$ to $\overline{\Omega}$.
(ii) $\mathcal{R}^{**} = (\mathcal{R}^*)^* = \mathcal{R}$.
(iii) If $x_0 \in \overline{\Omega}$ and $m_0 \in \overline{\Omega^*}$, then $x_0 \in \mathcal{N}_{\mathcal{R}^*}(m_0)$ iff $m_0 \in \mathcal{N}_{\mathcal{R}}(x_0)$.

Proof. Given $m_0 \in \overline{\Omega^*}$, $\rho(x)(1-\kappa x \cdot m_0)$ must attain the maximum over $\overline{\Omega}$ at some $x_0 \in \overline{\Omega}$. Then $\rho^*(m_0) = 1/[\rho(x_0)(1 - \kappa x_0 \cdot m_0)]$. We always have

$$\rho^*(m) = \inf_{x \in \overline{\Omega}} \frac{1}{\rho(x)(1 - \kappa m \cdot x)} \le \frac{1}{\rho(x_0)(1 - \kappa x_0 \cdot m)}, \qquad \forall m \in \overline{\Omega^*}. \tag{4}$$

Hence $E(x_0, 1/\rho(x_0))$ is a supporting semi-ellipsoid to $\mathcal{R}^*$ at $\rho^*(m_0)m_0$. Thus, (i) is proved.

To prove (ii), from the definitions of Legendre transform and refractor mapping we have

$$\rho(x_0)\,\rho^*(m_0) = \frac{1}{1 - \kappa m_0 \cdot x_0} \quad \text{for } m_0 \in \mathcal{N}_{\mathcal{R}}(x_0). \tag{5}$$

For $x_0 \in \overline{\Omega}$, there exists $m_0 \in \mathcal{N}_{\mathcal{R}}(x_0)$ and so from (5) $\rho^*(m_0) = \dfrac{1/\rho(x_0)}{1 - \kappa x_0 \cdot m_0}$. By (4), $\rho^*(m)(1 - k x_0 \cdot m)$ attains the maximum $1/\rho(x_0)$ at m_0. Thus,

$$\rho^{**}(x_0) = \inf_{m \in \overline{\Omega}^*} \frac{1}{\rho^*(m)(1 - k x_0 \cdot m)} = \frac{1}{\rho(x_0)^{-1}}.$$

To prove (iii), we get from the proof of (ii) that if $m_0 \in \mathcal{N}_{\mathcal{R}}(x_0)$, then the semi-ellipsoid $E(x_0, 1/\rho(x_0))$ supports $\mathcal{R}^*$ at $\rho^*(m_0)m_0$ and so $x_0 \in \mathcal{N}_{\mathcal{R}^*}(m_0)$. On the other hand, if $x_0 \in \mathcal{N}_{\mathcal{R}^*}(m_0)$, we get that $m_0 \in \mathcal{N}_{\mathcal{R}^{**}}(x_0)$, and since $\mathcal{R}^{**} = \mathcal{R}$, $m_0 \in \mathcal{N}_{\mathcal{R}}(x_0)$. $\qquad \square$

The next two lemmas discuss the refractor measure.

Lemma 3.12. $\mathcal{C} = \{F \subset \overline{\Omega}^* : \mathcal{T}_{\mathcal{R}}(F) \text{ is Lebesgue measurable}\}$ *is a σ-algebra containing all Borel sets in $\overline{\Omega}^*$.*

Proof. Obviously, $\mathcal{T}_{\mathcal{R}}(\emptyset) = \emptyset$ and $\mathcal{T}_{\mathcal{R}}(\overline{\Omega}^*) = \overline{\Omega}$. Since $\mathcal{T}_{\mathcal{R}}(\cup_{i=1}^{\infty} F_i) = \cup_{i=1}^{\infty} \mathcal{T}_{\mathcal{R}}(F_i)$, $\mathcal{C}$ is closed under countable unions. Clearly for $F \subset \overline{\Omega}^*$

$$\mathcal{T}_{\mathcal{R}}(F^c) = \{x \in \overline{\Omega} : \mathcal{N}_{\mathcal{R}}(x) \cap F^c \neq \emptyset\}$$

$$= \{x \in \overline{\Omega} : \mathcal{N}_{\mathcal{R}}(x) \cap F = \emptyset\} \cup \{x \in \overline{\Omega} : \mathcal{N}_{\mathcal{R}}(x) \cap F^c \neq \emptyset, \mathcal{N}_{\mathcal{R}}(x) \cap F \neq \emptyset\}$$

$$= [\mathcal{T}_{\mathcal{R}}(F)]^c \cup [\mathcal{T}_{\mathcal{R}}(F^c) \cap \mathcal{T}_{\mathcal{R}}(F)]. \tag{6}$$

If $x \in \mathcal{T}_{\mathcal{R}}(F^c) \cap \mathcal{T}_{\mathcal{R}}(F) \cap \Omega$, then $\mathcal{R}$ parameterized by ρ has two distinct supporting semi-ellipsoids $E(m_1, b_1)$ and $E(m_2, b_2)$ at $\rho(x)x$. We show that $\rho(x)x$ is a singular point of $\mathcal{R}$. Otherwise, if $\mathcal{R}$ has the tangent hyperplane Π at $\rho(x)x$, then Π must coincide both with the tangent hyperplane of $E(m_1, b_1)$ and that of $E(m_2, b_2)$ at $\rho(x)x$. It follows from the Snell law that $m_1 = m_2$. Therefore, the area measure of $\mathcal{T}_{\mathcal{R}}(F^c) \cap \mathcal{T}_{\mathcal{R}}(F)$ is 0. So $\mathcal{C}$ is closed under complements, and we have proved that $\mathcal{C}$ is a σ-algebra.

To prove that $\mathcal{C}$ contains all Borel subsets, it suffices to show that $\mathcal{T}_{\mathcal{R}}(K)$ is compact if $K \subset \overline{\Omega}^*$ is compact. Let $x_i \in \mathcal{T}_{\mathcal{R}}(K)$ for $i \geq 1$. There exists $m_i \in \mathcal{N}_{\mathcal{R}}(x_i) \cap K$. Let $E(m_i, b_i)$ be the supporting semi-ellipsoid to $\mathcal{R}$ at $\rho(x_i)x_i$. We have

$$\rho(x)(1 - \kappa m_i \cdot x) \leq b_i \quad \text{for } x \in \overline{\Omega}, \tag{7}$$

where the equality in (7) occurs at $x = x_i$. Assume that $a_1 \leq \rho(x) \leq a_2$ on $\overline{\Omega}$ for some constants $a_2 \geq a_1 > 0$. By (7) and (1), $a_1(1 - \kappa) \leq b_i \leq a_2(1 - \kappa^2)$. Assume through subsequence that $x_i \longrightarrow x_0$, $m_i \longrightarrow m_0 \in K$, $b_i \longrightarrow b_0$, as $i \longrightarrow \infty$. By taking limit in (7), one obtains that the semi-ellipsoid $E(m_0, b_0)$ supports $\mathcal{R}$ at $\rho(x_0)x_0$ and $x_0 \in \mathcal{T}_{\mathcal{R}}(m_0)$. This proves $\mathcal{T}_{\mathcal{R}}(K)$ is compact.

To show that $\mathcal{C}$ is closed by complements, it is enough to notice the formula that

$$\mathcal{T}_{\mathcal{R}}\left(\Omega^* \setminus F\right) = \left[\mathcal{T}_{\mathcal{R}}(\Omega^*) \setminus \mathcal{T}_{\mathcal{R}}(F)\right] \cup \left[\mathcal{T}_{\mathcal{R}}(\Omega^* \setminus F) \cap \mathcal{T}_{\mathcal{R}}(F)\right].$$

$\square$

Lemma 3.13. *Given a nonnegative $f \in L^1(\overline{\Omega})$, the set function*

$$\mathcal{M}_{\mathcal{R},f}(F) = \int_{\mathcal{T}_{\mathcal{R}}(F)} f \, dx$$

is a finite Borel measure defined on $\mathcal{C}$ and is called the refractor measure associated with $\mathcal{R}$ and f.

Proof. Let $\{F_i\}_{i=1}^{\infty}$ be a sequence of pairwise disjoint sets in $\mathcal{C}$. Let $H_1 = \mathcal{T}_{\mathcal{R}}(F_1)$, and $H_k = \mathcal{T}_{\mathcal{R}}(F_k) \setminus \cup_{i=1}^{k-1} \mathcal{T}_{\mathcal{R}}(F_i)$, for $k \geq 2$. Since $H_i \cap H_j = \emptyset$ for $i \neq j$ and $\cup_{k=1}^{\infty} H_k = \cup_{k=1}^{\infty} \mathcal{T}_{\mathcal{R}}(F_k)$, it is easy to get

$$\mathcal{M}_{\mathcal{R},f}(\cup_{k=1}^{\infty} F_k) = \int_{\cup_{k=1}^{\infty} H_k} f \, dx = \sum_{k=1}^{\infty} \int_{H_k} f \, dx.$$

Observe that $\mathcal{T}_{\mathcal{R}}(F_k) \setminus H_k = \mathcal{T}_{\mathcal{R}}(F_k) \cap (\cup_{i=1}^{k-1} \mathcal{T}_{\mathcal{R}}(F_i))$ is a subset of the singular set of $\mathcal{R}$ and has area measure 0 for $k \geq 2$. Therefore, $\int_{H_k} f \, dx = \mathcal{M}_{\mathcal{R},f}(F_k)$ and the σ-additivity of $\mathcal{M}_{\mathcal{R},f}$ follows. $\square$

The notion of weak solutions is introduced through the conservation of energy.

Definition 3.14. A refractor $\mathcal{R}$ is a weak solution of the refractor problem for the case $\kappa < 1$ with emitting illumination intensity $f(x)$ on $\overline{\Omega}$ and prescribed refracted illumination intensity μ on $\overline{\Omega}^*$ if for any Borel set $F \subset \overline{\Omega}^*$

$$\mathcal{M}_{\mathcal{R},f}(F) = \int_{\mathcal{T}_{\mathcal{R}}(F)} f \, dx = \mu(F). \tag{8}$$

3.3   Solution of the Refractor Problem

We introduce the cost

$$c(x, m) = \log \frac{1}{1 - \kappa \, x \cdot m}$$

for $x \in \Omega$ and $m \in \Omega^*$ where we assume $\Omega \cdot \Omega^* \geq \kappa$. From Definitions 3.2 and 3.8, $\mathcal{R} = \{\rho(x)x : x \in \overline{\Omega}\}$ is a refractor iff $\log \rho$ is c-concave. Using Definitions 3.3 and 3.9 we get that

$$\mathcal{N}_{c,\phi}(x) = \mathcal{N}_{\mathcal{R}}(x), \qquad \mathcal{R} = \{\rho(x)x : x \in \Omega\}, \qquad \rho(z) = e^{\phi(z)}.$$

Furthermore, $\log \rho^* = (\log \rho)^c$, $\log \rho = (\log \rho^*)_c$ by Remark (3) after Definition 3.2, and $\mathcal{N}_{\mathcal{R}}(x_0) = \mathcal{N}_{c,\log \rho}(x_0)$ by (5). By the Snell law and Lemma 3.12, $c(x,m)$ satisfies (5). From the definitions, $\mathcal{R}$ is a weak solution of the refractor problem iff $\log \rho$ is c-concave and $\mathcal{N}_{c,\log \rho}$ is a measure preserving mapping from $f(x)dx$ to μ.

By Lemma 3.5, there exists a c-concave $\phi(x)$ such that (ϕ, ϕ^c) maximizes

$$I(u, v) = \int_{\overline{\Omega}} u f \, dx + \int_{\overline{\Omega^*}} v \, d\mu(m)$$

in $\mathcal{K} = \{(u, v) \in C(\overline{\Omega}) \times C(\overline{\Omega^*}) : u(x) + v(m) \leq c(x, m), \text{ for } x \in \overline{\Omega}, m \in \overline{\Omega^*}\}$. Then by Lemma 3.4, $\mathcal{N}_{c,\phi}(x)$ is a measure preserving mapping from $f dx$ to μ. Therefore, $\mathcal{R} = \{e^{\phi(x)}x : x \in \overline{\Omega}\}$ is a weak solution of the refractor problem.

It remains to prove the uniqueness of solutions up to dilations. Let $\mathcal{R}_i = \{\rho_i(x)x : x \in \overline{\Omega}\}$, $i = 1, 2$, be two weak solutions of the refractor problem. Obviously, $\mathcal{N}_{c,\log \rho_i}$ have the converging property (C) stated before Lemma 3.7. It follows from Lemmas 3.4, 3.6 and 3.7 that $\mathcal{N}_{c,\log \rho_1}(x) = \mathcal{N}_{c,\log \rho_2}(x)$ a.e. on Ω. That is, $\mathcal{N}_{\mathcal{R}_1}(x) = \mathcal{N}_{\mathcal{R}_2}(x)$ a.e. on Ω. From the Snell law $\nu_i(x) = \dfrac{x - \kappa \mathcal{N}_{\mathcal{R}_i}(x)}{|x - \kappa \mathcal{N}_{\mathcal{R}_i}(x)|}$ is the unit normal to $\mathcal{R}_i$ towards medium II at $\rho_i(x)x$ where $\mathcal{R}_i$ is differentiable. So $\nu_1(x) = \nu_2(x)$ a.e. and consequently $\rho_1(x) = C \rho_2(x)$ for some $C > 0$.

4 Solution to the Refractor Problem for $\kappa < 1$ with the Minkowski Method

In this section we solve the refractor problem using a method having its roots in the Minkowski method from convex analysis, [13, Sect. 7.1]. In addition, to the definitions and lemmas from Sect. 3.2, we have the following.

Lemma 4.1. *We have for a refractor $\mathcal{R}$ that*

(i) $[\mathcal{T}_{\mathcal{R}}(F)]^c \subset \mathcal{T}_{\mathcal{R}}(F^c)$ *for all $F \subset \overline{\Omega^*}$, with equality except for a set of measure zero.*

(ii) *The set $\mathcal{C} = \{F \subset \overline{\Omega^*} : \mathcal{T}_{\mathcal{R}}(F) \text{ is Lebesgue measurable}\}$ is a σ-algebra containing all Borel sets in $\overline{\Omega^*}$.*

Lemma 4.2. *Let $\mathcal{R}_j = \{\rho_j(x)x : x \in \overline{\Omega}\}$, $j \geq 1$ be refractors from $\overline{\Omega}$ to $\overline{\Omega^*}$. Suppose that $0 < a_1 \leq \rho_j \leq a_2$ and $\rho_j \to \rho$ uniformly on $\overline{\Omega}$. Then:*

(i) $\mathcal{R} := \{\rho(x)x : x \in \overline{\Omega}\}$ *is a refractor from* $\overline{\Omega}$ *to* $\overline{\Omega^*}$.

(ii) *For any compact set* $K \subset \overline{\Omega^*}$

$$\limsup_{j \to \infty} \mathcal{T}_{\mathcal{R}_j}(K) \subset \mathcal{T}_{\mathcal{R}}(K).$$

(iii) *For any open set* $G \subset \overline{\Omega^*}$,

$$\mathcal{T}_{\mathcal{R}}(G) \subset \liminf_{j \to \infty} \mathcal{T}_{\mathcal{R}_j}(G) \cup S,$$

where S *is the singular set of* $\mathcal{R}$.

Proof. (i) Obviously $\rho \in C(\overline{\Omega})$ and $\rho > 0$. Fix $x_o \in \overline{\Omega}$. Then there exist $m_j \in \overline{\Omega^*}$ and $b_j > 0$ such that $E(m_j, b_j)$ supports $\mathcal{R}_j$ at $\rho(x_o)x_o$ and thus

$$\rho_j(x_o) = \frac{b_j}{1 - \kappa m_j \cdot x_o} \quad \text{and} \quad \rho_j(x) \le \frac{b_j}{1 - \kappa m_j \cdot x}$$

for all $x \in \overline{\Omega}$. Consequently

$$\frac{b_j}{1 - \kappa m_j \cdot x_o} \le a_2 \quad \text{and} \quad a_1 \le \frac{b_j}{1 - \kappa m_j \cdot x}$$

for all j and therefore

$$a_1(1 - \kappa) \le b_j \le a_2$$

for all j. If need be by passing to a subsequence we obtain m_o and b_o such that $m_j \to m_o \in \overline{\Omega^*}$ and $b_j \to b_o$. We claim $E(m_o, b_o)$ supports $\mathcal{R}$ at $\rho(x_o)x_o$. Indeed

$$\rho(x_o) = \lim_j \rho_j(x_o) = \lim_j \frac{b_j}{1 - \kappa m_j \cdot x_o} = \frac{b_o}{1 - \kappa m_o \cdot x_o}$$

and

$$\rho(x) = \lim_j \rho_j(x) \le \lim_j \frac{b_j}{1 - \kappa m_j \cdot x} = \frac{b_o}{1 - \kappa m_o \cdot x}$$

for all $x \in \overline{\Omega}$. Thus $\mathcal{R}$ is a refractor.

(ii) Let $x_o \in \limsup \mathcal{T}_{\mathcal{R}_j}(K)$. Without loss of generality assume that $x_o \in \mathcal{T}_{\mathcal{R}_j}(K)$ for all $j \ge 1$. Then there exist $m_j \in \mathcal{N}_{\mathcal{R}_j}(x_o) \cap K$ and b_j such that

$$\rho_j(x_o) = \frac{b_j}{1 - \kappa m_j \cdot x_o} \quad \text{and} \quad \rho_j(x) \le \frac{b_j}{1 - \kappa m_j \cdot x}$$

for all $x \in \overline{\Omega}$. As in the proof of (i) we may assume that $m_j \to m_o \in K$ and $b_j \to b_o$ and conclude that $E(m_o, b_o)$ supports $\mathcal{R}$ at $\rho(x_o)x_o$, proving that $x_o \in T_{\mathcal{R}}(m_o)$. Hence $x_o \in T_{\mathcal{R}}(K)$.

(iii) Let G be an open subset of $\overline{\Omega}^*$. By (ii) $\limsup T_{\mathcal{R}_j}(G^c) \subset T_{\mathcal{R}}(G^c)$ as G^c is compact. Also

$$\limsup_{j \to \infty}[T_{\mathcal{R}_j}(G)]^c \subset \limsup_{j \to \infty}\{[T_{\mathcal{R}_j}(G)]^c \cup [T_{\mathcal{R}_j}(G) \cap T_{\mathcal{R}_j}(G^c)]\} \quad (1)$$

and by Lemma 4.1 the right hand side of (1) is equal to $\limsup_{j \to \infty} T_{\mathcal{R}_j}(G^c)$. By (ii) we will then have

$$\limsup_{j \to \infty}[T_{\mathcal{R}_j}(G)]^c \subset T_{\mathcal{R}}(G^c) = \{[T_{\mathcal{R}}(G)]^c \cup [T_{\mathcal{R}}(G) \cap T_{\mathcal{R}}(G^c)]\}.$$

Taking complements we obtain

$$\{\limsup_{j \to \infty}[T_{\mathcal{R}_j}(G)]^c\}^c \supset [T_{\mathcal{R}}(G)] \cap [T_{\mathcal{R}}(G) \cap T_{\mathcal{R}}(G^c)]^c.$$

Consequently

$$\liminf_{j \to \infty} T_{\mathcal{R}_j}(G) \supset [T_{\mathcal{R}}(G)] \cap [T_{\mathcal{R}}(G) \cap T_{\mathcal{R}}(G^c)]^c$$

and thus

$$[[T_{\mathcal{R}}(G)] \cap [T_{\mathcal{R}}(G) \cap T_{\mathcal{R}}(G^c)]^c] \cup S \subset \liminf_{j \to \infty} T_{\mathcal{R}_j}(G) \cup S.$$

But $T_{\mathcal{R}}(G) \cap T_{\mathcal{R}}(G^c) \subset S$. Thus

$$T_{\mathcal{R}}(G) \subset T_{\mathcal{R}}(G) \cup S \subset \liminf_{j \to \infty} T_{\mathcal{R}_j}(G) \cup S$$

as required.

$\square$

Remark 4.3 (Invariance by dilations). Suppose that $\mathcal{R}$ is a refractor weak solution in the sense of Definition 3.14 with intensities f, μ and defined by $\rho(x)x$ for $x \in \bar{\Omega}$. Then for each $\alpha > 0$, the refractor $\alpha \mathcal{R}$ defined by $\alpha \rho(x)x$ for $x \in \bar{\Omega}$ is a weak solution in the sense of Definition 3.14 with the same intensities. In fact, $E(m, b)$ is a supporting ellipsoid to $\mathcal{R}$ at the point y if and only if $E(m, \alpha b)$ is a supporting ellipsoid to $\alpha \mathcal{R}$ at the point y. This means that $T_{\mathcal{R}}(m) = T_{\alpha \mathcal{R}}(m)$ for each $m \in \bar{\Omega}^*$.

4.1 Existence of Solutions in the Discrete Case

Theorem 4.4. *Let $f \in L^1(\Omega)$ with $\inf_\Omega f > 0$, $g_1, \cdots, g_N$ positive numbers, $m_1, \cdots, m_N \in \Omega^*$ distinct points, $N \geq 2$, with $x \cdot m_j \geq \kappa$ for all $x \in \Omega$ and $1 \leq j \leq N$. Let $\mu = \sum_{j=1}^{N} g_j \, \delta_{m_j}$, and assume the conservation of energy condition*

$$\int_{\Omega} f(x)\,dx = \mu(\Omega^*). \tag{2}$$

Then there exists a refractor $\mathcal{R}$ such that

(a) $\bar{\Omega} = \cup_{j=1}^{N} T_{\mathcal{R}}(m_j),$
(b) $\int_{T_{\mathcal{R}}(m_j)} f(x)\,dx = g_j$ *for* $1 \le j \le N.$

To prove the theorem, we prove first a sequence of lemmas.

Lemma 4.5. *Let*

$$W = \left\{ b = (1, b_2, \cdots, b_N) : b_j > 0,\ \mathcal{M}_{\mathcal{R}(b), f}(m_j) = \int_{T_{\mathcal{R}(b)}(m_j)} f(x)\,dx \le g_j,\ j = 2, \cdots, N \right\}, \tag{3}$$

where

$$\rho(x) = \mathcal{R}(b)(x) = \min_{1 \le j \le N} \frac{b_i}{1 - \kappa\, x \cdot m_j}. \tag{4}$$

Then, with the assumptions of Theorem 4.4, we have

(a) $W \ne \emptyset$
(b) *if* $b = (1, b_2, \cdots, b_N) \in W$, *then* $b_j > \dfrac{1}{1 + \kappa}$ *for* $j = 2, \cdots, N.$

Proof. (a) If for some $j \ne 1$, the semi-ellipsoid $E(m_j, b)$ supports $\mathcal{R}(b)$ at some $x \in \Omega$, then $\rho(z) \le \dfrac{b}{1 - \kappa\, z \cdot m_j}$ for all $z \in \Omega$, and $\rho(x) = \dfrac{b}{1 - \kappa\, x \cdot m_j}$. Since $x \cdot m_j \ge \kappa$, we have

$$\frac{b}{1 - \kappa^2} \le \frac{b}{1 - \kappa\, x \cdot m_j} = \rho(x) \le \frac{1}{1 - \kappa\, x \cdot m_1} \le \frac{1}{1 - \kappa},$$

and so $b \le 1 + \kappa$. Therefore, if $b_i > 1 + \kappa$ for $2 \le i \le N$, then $E(m_i, b_i)$ cannot be a supporting ellipsoid to $\mathcal{R}(b)$ at any $x \in \Omega$. On the other hand, if $x \in T_{\mathcal{R}}(m_j)$, then $m_j \in \mathcal{N}_{\mathcal{R}}(x)$ and if x is not a singular point of $\mathcal{R}(b)$ there is a unique ellipsoid $E(m_j, b)$ supporting $\mathcal{R}$ at x. But from the definition of $\mathcal{R}$ there is an ellipsoid $E(m_k, b_k)$ that supports $\mathcal{R}$ at x, and so $E(m_k, b_k) = E(m_j, b)$, i.e., $k = j$ and $b = b_j$. Consequently the set $T_{\mathcal{R}}(m_j)$ is contained in the set of singular points and therefore has measure zero. So $\mathcal{M}_{\mathcal{R}(b), f}(m_j) = 0 < g_j$ for $j = 2, \cdots, N$ and so any point $b = (1, b_2, \cdots, b_N) \in W$ as long as $b_i > 1 + \kappa$ for $i = 2, \cdots, N$.

(b) First notice that if $E(m_j, b_j)$ and $E(m_k, b_k)$ support $\mathcal{R}(b)$ at x_0, then x_0 is a singular point. And therefore, $|T_{\mathcal{R}}(m_j) \cap T_{\mathcal{R}}(m_k)| = 0$ for $k \ne j$.

Claim 1. If $b \in W$, then $g_1 \le \mathcal{M}_{\mathcal{R}(b), f}(m_1)$.

Indeed,

$$\sum_{i=1}^{N} \mathcal{M}_{\mathcal{R}(b),f}(m_i) = \sum_{i=1}^{N} \int_{\mathcal{T}_{\mathcal{R}(b)}(m_i)} f(x)\,dx = \int_{\cup_{i=1}^{N}\mathcal{T}_{\mathcal{R}(b)}(m_i)} f(x)\,dx = \int_{\Omega} f(x)\,dx = \mu(\Omega^*) = \sum_{i=1}^{N} g_i,$$

from (2). Hence

$$g_1 - \mathcal{M}_{\mathcal{R}(b),f}(m_1) + \sum_{i=2}^{N} \left(g_i - \mathcal{M}_{\mathcal{R}(b),f}(m_i) \right) = 0.$$

If $b \in W$, then $g_i - \mathcal{M}_{\mathcal{R}(b),f}(m_i) \geq 0$ for $i = 2, \cdots, N$, and Claim 1 follows.

Claim 2. For each $b \in W$, $\mathcal{T}_{\mathcal{R}(b)}(m_1) \cap \left(\cup_{i=2}^{N}\mathcal{T}_{\mathcal{R}(b)}(m_i) \right)^c \neq \emptyset$.

Otherwise, $\mathcal{T}_{\mathcal{R}(b)}(m_1) \subset \cup_{i=2}^{N}\mathcal{T}_{\mathcal{R}(b)}(m_i)$ which means that each point in $\mathcal{T}_{\mathcal{R}(b)}(m_1)$ is singular, and therefore $|\mathcal{T}_{\mathcal{R}(b)}(m_1)| = 0$. This contradicts Claim 1, since $g_1 > 0$.

Therefore, if $b \in W$, then we can pick $x_0 \in \mathcal{T}_{\mathcal{R}(b)}(m_1) \cap \left(\cup_{i=2}^{N}\mathcal{T}_{\mathcal{R}(b)}(m_i) \right)^c$ and so

$$\rho(x_0) = \frac{1}{1 - \kappa\, x_0 \cdot m_1} < \frac{b_i}{1 - \kappa\, x_0 \cdot m_i}, \qquad i = 2, \cdots, N$$

so

$$b_i > \frac{1 - \kappa\, x_0 \cdot m_i}{1 - \kappa\, x_0 \cdot m_1} \geq \frac{1 - \kappa\, x_0 \cdot m_i}{1 - \kappa^2} \geq \frac{1 - \kappa}{1 - \kappa^2} = \frac{1}{1 + \kappa}.$$

$\square$

Lemma 4.6. *If $b_j = (b_1^j, \cdots, b_N^j) \to b_0 = (b_1^0, \cdots, b_N^0)$ as $j \to \infty$, then $\rho_j = \mathcal{R}(b_j) \to \rho_0 = \mathcal{R}(b_0)$ uniformly in $\bar{\Omega}$ as $j \to \infty$.*

Proof. Given $y \in \bar{\Omega}$, there exists $1 \leq \ell \leq N$ such that $\rho_0(y) = \dfrac{b_\ell^0}{1 - \kappa\, y \cdot m_\ell}$. Hence

$$\rho_j(y) - \rho_0(y) \leq \frac{b_\ell^j}{1 - \kappa\, y \cdot m_\ell} - \frac{b_\ell^0}{1 - \kappa\, y \cdot m_\ell} \leq \frac{|b_\ell^j - b_\ell^0|}{1 - \kappa\, y \cdot m_\ell} \leq \frac{|b_\ell^j - b_\ell^0|}{1 - \kappa} \to 0,$$

as $j \to \infty$.

$\square$

Lemma 4.7. *Let $\delta > 0$ and the region $R_\delta = \{(1, b_2, \cdots, b_N) : b_j \geq \delta, 2 \leq j \leq N\}$. The functions $G_{\mathcal{R}(b)}(m_i) := \mathcal{M}_{\mathcal{R}(b)}(m_i)$ are continuous for $b \in R_\delta$ for $i = 1, 2, \cdots, N$.*

Proof. Let $b_j = (1, b_2^j, \cdots, b_N^j) \in R_\delta$ with $b_j \to b_0$ as $j \to \infty$. By Lemma 4.6, $\rho_j \to \rho_0$ uniformly in $\bar{\Omega}$. Given $x \in \bar{\Omega}$, we have $\rho_j(x) = \dfrac{b_\ell^j}{1 - \kappa\, x \cdot m_\ell}$ for some $1 \leq \ell \leq N$ and so $\rho_j(x) \geq \dfrac{\min\{1,\delta\}}{1+\kappa}$. On the other hand, $\rho_j(x) =$

$\displaystyle\min_{1\leq\ell\leq N} \dfrac{b_\ell^j}{1 - \kappa\, x \cdot m_\ell} \leq \dfrac{1}{1 - \kappa\, x \cdot m_1} \leq \dfrac{1}{1-\kappa}$. Therefore

$$\frac{\min\{1,\delta\}}{1+\kappa} \leq \rho_j(x) \leq \frac{1}{1-\kappa}$$

for all $x \in \bar{\Omega}$ and for all j.

Let us fix $1 \leq i \leq N$. Let $G \subset \bar{\Omega}^*$ be a neighborhood of m_i such that $m_\ell \notin G$ for $\ell \neq i$. If $x_0 \in \mathcal{T}_{\mathcal{R}(b_j)}(G)$ and x_0 is not a singular point, then there exists a unique $m \in G$ and $b > 0$ such that $\rho_j(x_0) = \dfrac{b}{1 - \kappa\, x_0 \cdot m}$ and $\rho_j(z) \leq \dfrac{b}{1 - \kappa\, z \cdot m}$ for all $x \in \bar{\Omega}$. From the definition of $\mathcal{R}(b_j)$ and since x_0 is not singular, $m = m_\ell$. Since $m \in G$, we get $m = m_i$. Therefore

$$\mathcal{T}_{\mathcal{R}(b_j)}(G) \subset \mathcal{T}_{\mathcal{R}(b_j)}(m_i) \cup S,$$

where S is the set of singular points. By Lemma 4.2

$$\mathcal{T}_{\mathcal{R}(b_0)}(G) \subset \liminf_{j\to\infty} \mathcal{T}_{\mathcal{R}(b_j)}(G) \cup S,$$

and we therefore obtain

$$\mathcal{T}_{\mathcal{R}(b_0)}(G) \subset \liminf_{j\to\infty} \mathcal{T}_{\mathcal{R}(b_j)}(m_i) \cup S.$$

Thus

$$\int_{\mathcal{T}_{\mathcal{R}(b_0)}(m_i)} f(x)\,dx \leq \int_{\mathcal{T}_{\mathcal{R}(b_0)}(G)} f(x)\,dx \leq \int_{\liminf_{j\to\infty}\mathcal{T}_{\mathcal{R}(b_j)}(m_i)} f(x)\,dx$$

$$\leq \liminf_{j\to\infty} \int_{\mathcal{T}_{\mathcal{R}(b_j)}(m_i)} f(x)\,dx \qquad \text{by Fatou.}$$

We next prove that

$$\limsup_{j\to\infty} \int_{\mathcal{T}_{\mathcal{R}(b_j)}(m_i)} f(x)\,dx \leq \int_{\mathcal{T}_{\mathcal{R}(b_0)}(m_i)} f(x)\,dx.$$

By Lemma 4.2

$$\limsup_{j\to\infty} \mathcal{T}_{\mathcal{R}(b_j)}(K) \subset \mathcal{T}_{\mathcal{R}(b_0)}(K)$$

for each K compact. Hence

$$\int_{\limsup_{j\to\infty}\mathcal{T}_{\mathcal{R}(b_j)}(m_i)} f(x)\,dx \le \int_{\mathcal{T}_{\mathcal{R}(b_0)}(m_i)} f(x)\,dx.$$

By reverse Fatou we have

$$\limsup_{j\to\infty}\int_{\mathcal{T}_{\mathcal{R}(b_j)}(m_i)} f(x)\,dx \le \int_{\limsup_{j\to\infty}\mathcal{T}_{\mathcal{R}(b_j)}(m_i)} f(x)\,dx$$

and therefore the lemma is proved. $\square$

Proof of Theorem 4.4. Fix $\tilde{b} = (1, \tilde{b}_2, \cdots, \tilde{b}_N) \in W$ and let

$$\tilde{W} = \{b = (1, b_2, \cdots, b_N) \in W : b_j \le \tilde{b}_j,\ j = 2, \cdots, N\}.$$

$\tilde{W}$ is compact. Let $d : \tilde{W} \to \mathbb{R}$ be given by $d(b) = 1 + b_2 + \cdots + b_N$; d attains its minimum value in $\tilde{W}$ at a point $b^* = (1, b_2^*, \cdots, b_N^*)$ (notice that the minimum is strictly positive by Lemma 4.5(b)). We prove that $\mathcal{R}(b^*)$ is the refractor that solves the problem. By conservation of energy it is enough to show that $\int_{\mathcal{T}_{\mathcal{R}(b^*)}(m_j)} f(x)\,dx = g_j$ for $j = 2, \cdots, N$. Since $b^* \in W$, we have $\int_{\mathcal{T}_{\mathcal{R}(b^*)}(m_j)} f(x)\,dx \le g_j$ for $j = 2, \cdots, N$. Suppose by contradiction that this inequality is strict for some j, suppose for example that

$$\int_{\mathcal{T}_{\mathcal{R}(b^*)}(m_2)} f(x)\,dx < g_2. \tag{5}$$

Let $0 < \lambda < 1$ and $b_\lambda^* = (1, \lambda b_2^*, b_3^*, \cdots, b_N^*)$.

   We claim that

$$\mathcal{T}_{\mathcal{R}(b_\lambda^*)}(m_i) \setminus \text{ set of measure zero} \subset \mathcal{T}_{\mathcal{R}(b^*)}(m_i) \tag{6}$$

for $i = 3, 4, \cdots, N$ and all $0 < \lambda < 1$. Indeed, if $x_0 \in \mathcal{T}_{\mathcal{R}(b_\lambda^*)}(m_i)$ is not a singular point of $\mathcal{R}(b_\lambda^*)$, then there is a unique ellipsoid $\dfrac{a}{1 - \kappa\, x \cdot m_i}$ that supports $\mathcal{R}(b_\lambda^*)$ at x_0 for some $a > 0$. Since $\mathcal{R}(b_\lambda^*)(x) = \min_{1 \le i \le N} \dfrac{(b_\lambda^*)_i}{1 - \kappa\, x \cdot m_i}$, there exists $1 \le j \le N$ such that $\mathcal{R}(b_\lambda^*)(x_0) = \dfrac{(b_\lambda^*)_j}{1 - \kappa\, x \cdot m_j}$. That is, the ellipsoid $\dfrac{(b_\lambda^*)_j}{1 - \kappa\, x \cdot m_j}$ supports $\mathcal{R}(b_\lambda^*)$ at x_0. Therefore, $\dfrac{a}{1 - \kappa\, x \cdot m_i} = \dfrac{(b_\lambda^*)_j}{1 - \kappa\, x \cdot m_j}$ implying $j = i$ and so $a = (b_\lambda^*)_i = b_i^*$. Since $\dfrac{b_i^*}{1 - \kappa\, x \cdot m_i} \ge \mathcal{R}(b^*)(x) \ge \mathcal{R}(b_\lambda^*)(x)$ for all x, it follows that $\dfrac{b_i^*}{1 - \kappa\, x \cdot m_i}$ is a supporting ellipsoid to $\mathcal{R}(b^*)$ at x_0. This proves the claim.

This implies that

$$\int_{\mathcal{T}_{\mathcal{R}(b_\lambda^*)}(m_i)} f(x)\,dx \le \int_{\mathcal{T}_{\mathcal{R}(b^*)}(m_i)} f(x)\,dx \le g_i$$

for $i = 3, 4, \cdots, N$ and all $0 < \lambda < 1$.

Finally from Lemma 4.7, inequality (5) holds for all λ sufficiently close to one, and therefore the point $b_\lambda^* \in \tilde{W}$ for all λ close to one. This is a contradiction because $d(b_\lambda^*) < d(b^*)$.   □

Remark 4.8. Notice that the solution in Theorem 4.4 has the form given by formula (4), where $b_1 = 1$ and $(1, b_2, \cdots, b_N) \in W$. So from Lemma 4.5(b), we have $b_i > 1/(1 + \kappa)$ for $i = 2, \cdots, N$. This implies that $\inf_{\bar{\Omega}} \rho(x) = \alpha > 0$.

4.2   Solution in the General Case

Lemma 4.9. *Let $\mathcal{R} = \{\rho(x)x : x \in \overline{\Omega}\}$ be a refractor from $\overline{\Omega}$ to $\overline{\Omega^*}$ such that $\inf_{x \in \overline{\Omega}} \rho(x) = 1$. Then there is a constant C, depending only on κ, such that*

$$\sup_{x \in \overline{\Omega}} \rho(x) \le C.$$

Proof. Suppose $\inf_{x \in \overline{\Omega}} \rho(x)$ is attained at $x_o \in \overline{\Omega}$, and let $E(m, b)$ be a supporting semi-ellipsoid to $\mathcal{R}$ at $\rho(x_0)x_0$. Then

$$1 = \rho(x_0) = \frac{b}{1 - \kappa\, m \cdot x_0} \quad \text{and} \quad \rho(x) \le \frac{b}{1 - \kappa\, m \cdot x} \quad \forall x \in \overline{\Omega}.$$

From the first equation we get that $b \le 1 + \kappa$, and using this in the inequality we obtain

$$\rho(x) \le \frac{1 + \kappa}{1 - \kappa^2} \quad \text{for all } x \in \overline{\Omega}$$

which proves the lemma.   □

Theorem 4.10. *Let $f \in L^1(\overline{\Omega})$ with $\inf_{\overline{\Omega}} f > 0$, and let μ be a Radon measure on $\overline{\Omega^*}$, such that*

$$\int_{\overline{\Omega}} f(x)\,dx = \mu(\overline{\Omega^*})$$

Then there exists a weak solution $\mathcal{R}$ of the refractor problem for the case $\kappa < 1$, with emitting illumination intensity f and prescribed refracted illumination intensity μ.

Proof. Fix $l \in \mathbf{N}, l \geq 2$. Partition $\overline{\Omega^*}$ into a finite number of disjoint Borel subsets $\omega_1^l, \ldots, \omega_{k_l}^l$ such that $\text{diam}(\omega_i^l) \leq \frac{1}{l}$. Choose points $m_i^l \in \omega_i^l$ and define a measure on $\overline{\Omega^*}$

$$\mu_l = \sum_{i=1}^{k_l} \mu(\omega_i^l) \delta_{m_i^l}.$$

Then

$$\mu_l(\overline{\Omega^*}) = \sum_{i=1}^{k_l} \mu(\omega_i^l) = \mu(\overline{\Omega^*}) = \int_{\overline{\Omega^*}} f(x)\, dx.$$

If $h \in C(\overline{\Omega^*})$, then

$$\int_{\overline{\Omega^*}} h\, d\mu_l - \int_{\overline{\Omega^*}} h\, d\mu = \sum_{i=1}^{k_l} \left(\int_{\omega_i^l} h(x)\, d\mu_l - \int_{\omega_i^l} h(x)\, d\mu \right)$$

$$= \sum_{i=1}^{k_l} \left(\int_{\omega_i^l} h(m_i^l)\, d\mu - \int_{\omega_i^l} h(x)\, d\mu \right)$$

$$= \sum_{i=1}^{k_l} \int_{\omega_i^l} (h(m_i^l) - h(x))\, d\mu.$$

Since $h \in C(\overline{\Omega^*})$ and $\text{diam}(\omega_i^l) < \frac{1}{l}$, we obtain that

$$\int_{\overline{\Omega^*}} h\, d\mu_l \to \int_{\overline{\Omega^*}} h\, d\mu \quad \text{as } l \to \infty$$

and hence μ_l *converges weakly* to μ.

By Theorem 4.4, let $\mathcal{R}_l = \{\rho_l(x)x : x \in \overline{\Omega}\}$ be the solution corresponding to μ_l, that is,

$$\mathcal{M}_{\mathcal{R}_l, f}(\omega) = \mu_l(\omega)$$

for every Borel subset ω of $\overline{\Omega^*}$. Notice that from Remark 4.8, $\inf_{\overline{\Omega}} \rho_l(x) = \alpha_l > 0$. In view of Remark 4.3, the refractor defined by the function $\frac{\rho_l(x)}{\alpha_l}$ solves the same problem and $\inf_{\overline{\Omega}} \frac{\rho_l(x)}{\alpha_l} = 1$. So normalizing ρ_l, we may assume that $\inf_{\overline{\Omega}} \rho_l(x) = 1$. Then by Lemma (4.9) there exists a uniform bound $C = C(\kappa)$ such that

$$\sup_{x \in \overline{\Omega}} \rho_l(x) \leq C \quad \text{for all } l \geq 1.$$

Also if $x_o, x_1 \in \overline{\Omega}$ and $E(m_o, b_o)$ is a supporting semi ellipsoid to $\mathcal{R}_l$ at $\rho_l(x_o)x_o$ then for $x_1 \in \overline{\Omega}$ we have

$$\rho_l(x_1) - \rho_l(x_o) \leq \frac{b_o}{1 - \kappa\, m_o \cdot x_1} - \frac{b_o}{1 - \kappa\, m_o \cdot x_o} = \frac{\kappa\, b_0\, m_o \cdot (x_1 - x_0)}{(1 - \kappa\, m_o \cdot x_1)(1 - \kappa\, m_o \cdot x_o)}$$

$$\leq \frac{\kappa\, b_0\, |m_o|\, |x_1 - x_0|}{(1 - \kappa\, m_o \cdot x_1)(1 - \kappa\, m_o \cdot x_o)} \leq C\, \frac{\kappa}{1 - \kappa} |x_1 - x_o|.$$

By changing the roles of x_o and x_1 we conclude that

$$|\rho_l(x_1) - \rho_l(x_o)| \leq C \frac{\kappa}{1 - \kappa} |x_1 - x_o| \text{ for all } l \geq 1.$$

Thus $\{\rho_l : l \geq 1\}$ is an equicontinuous family which is bounded uniformly. Then by *Arzelà-Ascoli* Theorem, if need be by taking subsequence, we have that $\rho_l \to \rho$ uniformly on $\overline{\Omega}$. By Lemma 4.2(i), $\mathcal{R} = \{\rho(x)x : x \in \overline{\Omega}\}$ is a refractor.

We claim that $\mathcal{M}_{\mathcal{R}_l, f}$ converges weakly to $\mathcal{M}_{\mathcal{R}, f}$. Indeed, if F is any closed subset of $\overline{\Omega}^*$ then by Lemma 4.2(ii) and reverse Fatou we have

$$\limsup_{l \to \infty} \mathcal{M}_{\mathcal{R}_l, f}(F) \leq \int_{\limsup_{l \to \infty} T_{\mathcal{R}_l}(F)} f(x)\, dx \leq \int_{T_{\mathcal{R}}(F)} f(x)\, dx = \mathcal{M}_{\mathcal{R}, f}(F).$$

Moreover for any open set $G \subset \overline{\Omega}^*$ we claim that

$$\mathcal{M}_{\mathcal{R}_l, f}(G) = \int_{T_{\mathcal{R}}(G)} f(x)\, dx \leq \liminf_{\ell \to \infty} \mathcal{M}_{\mathcal{R}_l, f}(G). \tag{7}$$

Indeed, from Lemma 4.2(iii) we have

$$\mathcal{M}_{\mathcal{R}, f}(G) = \int_{T_{\mathcal{R}}(G)} f(x)\, dx \leq \int_{\liminf_{\ell \to \infty} T_{\mathcal{R}_\ell}(G)} f(x)\, dx$$

$$= \int_{\bar{\Omega}^*} \liminf_{\ell \to \infty} \chi_{T_{\mathcal{R}_\ell}(G)}(x)\, f(x)dx$$

$$\leq \liminf_{\ell \to \infty} \int_{\bar{\Omega}^*} \chi_{T_{\mathcal{R}_\ell}(G)}(x)\, f(x)\, dx = \liminf_{\ell \to \infty} \mathcal{M}_{\mathcal{R}_l, f}(G),$$

by Fatou's lemma. Consequently $\mathcal{M}_{\mathcal{R}_l, f} \to \mathcal{M}_{\mathcal{R}, f}$ weakly.

Since $\mathcal{M}_{\mathcal{R}_l, f}(\omega) = \mu_l(\omega)$ for each Borel set ω, it follows by uniqueness of the weak limit that $\mathcal{M}_{\mathcal{R}, f} = \mu$.

$\square$

4.3   Uniqueness Discrete Case

With the method used in this section we show uniqueness when the target measure is discrete. Notice that uniqueness (up to dilations) was already proved with the mass transport approach in Sect. 3.3. But the method we describe here is applicable to the near field problem which is not a mass transport problem.

Recall that $(b_1, \cdots, b_N)$ are positive numbers, $m_1, \cdots, m_N$ are different points in the sphere S^{n-1}, and $\Omega \subset S^{n-1}$ with $\inf_{x \in \bar{\Omega}, 1 \leq j \leq N} x \cdot m_j \geq \kappa$. We let

$$\rho(x) = \min_{1 \leq i \leq N} \frac{b_i}{1 - \kappa x \cdot m_i},$$

and let $\mathcal{R} = \mathcal{R}(\mathbf{b}) = \{\rho(x)x : x \in \bar{\Omega}\}$.

Lemma 4.11. *Suppose that the set $\mathcal{T}_{\mathcal{R}}(m_j)$ has positive measure. If $x_0 \in \mathcal{T}_{\mathcal{R}}(m_j)$, then the semi-ellipsoid $E(m_j, b_j)$ supports $\mathcal{R}$ at the point x_0.*

Proof. We have $m_j \in \mathcal{N}_{\mathcal{R}}(x_0)$, that is, there exists a supporting semi-ellipsoid $E(m_j, b)$ to $\mathcal{R}$ at the point x_0 for some $b > 0$. We claim that $b = b_j$, and therefore $\dfrac{b_j}{1 - \kappa m_j \cdot x}$ supports $\mathcal{R}$ at x_0. Since $E(m_j, b)$ supports $\mathcal{R}$, we have $\rho(x) \leq \dfrac{b}{1 - \kappa x \cdot m_j}$ for all $x \in \bar{\Omega}$ with equality at $x = x_0$. Hence $\dfrac{b}{1 - \kappa x_0 \cdot m_j} \leq \dfrac{b_j}{1 - \kappa x_0 \cdot m_j}$, and so $b \leq b_j$. If $b = b_j$ we are done. If $b < b_j$, then

$$\rho(x) \leq \frac{b}{1 - \kappa x \cdot m_j} < \frac{b_j}{1 - \kappa x \cdot m_j}, \quad \forall x \in \bar{\Omega},$$

so $\rho(x) = \min_{k \neq j} \dfrac{b_k}{1 - \kappa m_k \cdot x}$, and therefore $\mathcal{R}$ cannot refract in the direction m_j (except on a set of directions with measure zero) and so $\mathcal{T}_{\mathcal{R}}(m_j)$ has measure zero, a contradiction. $\qquad\square$

Lemma 4.12. *Let $\mathcal{R}_b, \mathcal{R}_{b^*}$ be two solutions from Theorem 4.4, with $b = (b_1, \cdots, b_N)$, and $b^* = (b_1^*, \cdots, b_N^*)$. Assume that $f > 0$ a.e. in Ω. If $b_1^* \leq b_1$, then $b_i^* \leq b_i$ for all $1 \leq i \leq N$. In particular, if $b_1^* = b_1$, then $b_i^* = b_i$ for all $1 \leq i \leq N$.*

Proof. Let $J = \{j : b_j < b_j^*\}$ and $I = \{i : b_i \geq b_i^*\}$. Suppose by contradiction that $J \neq \emptyset$. We have $I \neq \emptyset$ since $1 \in I$. Fix $j \in J$, we have $\dfrac{b_j}{1 - \kappa z \cdot m_j} < \dfrac{b_j^*}{1 - \kappa z \cdot m_j}$ for all $z \in \bar{\Omega}$ since $b_j < b_j^*$. And also $\dfrac{b_i^*}{1 - \kappa z \cdot m_i} \leq \dfrac{b_i}{1 - \kappa z \cdot m_i}$ for all $i \in I$ and all $z \in \bar{\Omega}$. Fix $j \in J$ and let $x \in \mathcal{T}_{\mathcal{R}_{b^*}}(m_j)$. Since $\mathcal{R}_{b^*}$ is a solution to the discrete problem and $g_i > 0$ for all $1 \le i \le N$, we have that $\mathcal{T}_{\mathcal{R}_{b*}}(m_j)$ has positive measure. So from Lemma 4.11, the ellipsoid $\dfrac{b_j^*}{1 - \kappa m_j \cdot z}$ supports $\mathcal{R}_{b*}$ at the point x. Since the function defining $\mathcal{R}_{b*}$ is given by $\rho^*(z) = \min_{1 \le i \le N} \dfrac{b_i^*}{1 - \kappa m_i \cdot z}$, and $\rho^*(x) = \dfrac{b_j^*}{1 - \kappa m_j \cdot x}$, we therefore obtain

$$\frac{b_j}{1 - \kappa m_j \cdot x} < \frac{b_j^*}{1 - \kappa m_j \cdot x} \le \frac{b_i^*}{1 - \kappa m_i \cdot x} \le \frac{b_i}{1 - \kappa m_i \cdot x}, \quad \forall i \in I.$$

Hence by continuity, there exists N_x an open neighborhood of x such that

$$\frac{b_j}{1 - \kappa m_j \cdot y} < \frac{b_i}{1 - \kappa m_i \cdot y} \quad \forall i \in I \quad \forall y \in N_x.$$

Since the function defining $\mathcal{R}_b$ is $\rho(z) = \min_{1 \le i \le N} \dfrac{b_i}{1 - \kappa m_i \cdot z}$, we get for $y \in N_x$ that $\rho(y) = \min_{j \in J} \dfrac{b_j}{1 - \kappa m_j \cdot y}$, that is, $\rho(y) = \dfrac{b_{j'}}{1 - \kappa m_{j'} \cdot y}$ for some $j' \in J$ (depending also on y) which means that $\dfrac{b_{j'}}{1 - \kappa m_{j'} \cdot y}$ is a supporting ellipsoid to $\mathcal{R}_b$ at y. Therefore

$$N_x \subset \mathcal{T}_{\mathcal{R}_b}\left(\cup_{j \in J} m_j\right).$$

We then have that every point $x \in \mathcal{T}_{\mathcal{R}_{b*}}\left(\cup_{j \in J} m_j\right)$ has a neighborhood contained in $\mathcal{T}_{\mathcal{R}_b}\left(\cup_{j \in J} m_j\right)$, that is,

$$\mathcal{T}_{\mathcal{R}_{b*}}\left(\cup_{j \in J} m_j\right) \subset \left(\mathcal{T}_{\mathcal{R}_b}\left(\cup_{j \in J} m_j\right)\right)^\circ \ne \overline{\Omega}. \tag{8}$$

Since $\overline{\Omega}$ is connected and $\mathcal{T}_{\mathcal{R}_{b*}}\left(\cup_{j \in J} P_j\right)$ is closed, we get that $\left(\mathcal{T}_{\mathcal{R}_b}\left(\cup_{j \in J} m_j\right)\right)^\circ \setminus \mathcal{T}_{\mathcal{R}_{b*}}\left(\cup_{j \in J} m_j\right)$ is a non empty open set. This is a contradiction with the fact that

$$\int_{\mathcal{T}_{\mathcal{R}_b}(\cup_{j \in J} m_j)} f(x)\, dx = \sum_{j \in J} f_j = \int_{\mathcal{T}_{\mathcal{R}_{b*}}(\cup_{j \in J} m_j)} f(x)\, dx,$$

since $f > 0$ a.e.. $\qquad\qquad\qquad\qquad\qquad\qquad\qquad\qquad\qquad\qquad\qquad\qquad\square$

From the lemma we deduce the uniqueness up to dilations in the discrete case. Let $\lambda > 0$. Notice that if $E(m, b)$ is a supporting ellipsoid to the refractor $\mathcal{R}_\lambda$ with defining function $\lambda \rho(x)$ at x_0 if and only if $E(m, b/\lambda)$ is a supporting ellipsoid to the refractor $\mathcal{R}$ with defining function $\rho(x)$ at the point x_0. This implies that $\mathcal{N}_{\mathcal{R}_\lambda}(x_0) = \mathcal{N}_{\mathcal{R}}(x_0)$, and consequently $\mathcal{T}_{\mathcal{R}_\lambda}(m) = \mathcal{T}_{\mathcal{R}}(m)$. Therefore, if $\mathcal{R}$ is a refractor solving the problem in Theorem 4.4, then $\mathcal{R}_\lambda$ solves the same problem for any $\lambda > 0$. We now prove the uniqueness. Suppose $\mathcal{R}_b$ and $\mathcal{R}_{b^*}$ are two solutions as in Theorem 4.4. Pick λ such that $\lambda b_1 = b_1^*$. The refractor $\mathcal{R}_{\lambda b}$ is also a solution to Theorem 4.4, and by Lemma 4.12 we obtain that $\lambda b_i = b_i^*$ for all i. This means that $\mathcal{R}_b$ and $\mathcal{R}_{b^*}$ are multiples one of each other and we obtain the uniqueness.

5 Maxwell's Equations

The *electromagnetic field* (EM) is a physical field produced by electrically charged objects. It extends indefinitely throughout space and describes the electromagnetic interaction. The field propagates by electromagnetic radiation; in order of increasing energy (decreasing wavelength) electromagnetic radiation comprises: radio waves, microwaves, infrared, visible light, ultraviolet, X-rays, and gamma rays. The field (EM) can be viewed as the combination of an electric field $\mathbf{E}$ and a magnetic field $\mathbf{H}$, that is, these are three-dimensional vector fields that have a value defined at every point of space and time: $\mathbf{E} = \mathbf{E}(\mathbf{r}, t)$ and $\mathbf{H} = \mathbf{H}(\mathbf{r}, t)$, where $\mathbf{r}$ represents a point in 3-d space $\mathbf{r} = (x, y, z)$. The electric field is produced by stationary charges, and the magnetic field by moving charges (currents); these two are often described as the sources of the field. The way in which $\mathbf{E}$ and $\mathbf{H}$ interact is described by Maxwell's equations (see [14, p. 22]):

$$\nabla \cdot \mathbf{E} = \frac{\rho}{\epsilon_0}, \quad \text{Gauss's law} \tag{1}$$

$$\nabla \cdot \mathbf{H} = 0, \quad \text{Gauss's law for magnetism} \tag{2}$$

$$\nabla \times \mathbf{E} = -\frac{\partial \mathbf{H}}{\partial t}, \quad \text{Faraday's law} \tag{3}$$

$$\nabla \times \mathbf{H} = \mu_0 \mathbf{J} + \epsilon_0 \mu_0 \frac{\partial \mathbf{E}}{\partial t}, \quad \text{Ampère-Maxwell's law.} \tag{4}$$

Here

$\nabla = (\partial_x, \partial_y, \partial_z)$	the gradient
$\rho = \rho(\mathbf{r}, t)$	charge density
ϵ_0	permittivity of free space
$\mathbf{J} = \mathbf{J}(\mathbf{r}, t)$	current density vector
μ_0	permeability of free space

We have $c = 1/\sqrt{\epsilon_0 \mu_0}$, the speed of light in vacuum.

5.1   General Case

In several situations is necessary to consider a medium where the magnetic permeability $\mu = \mu(x, y, z)^3$ and the electric permittivity $\epsilon = \epsilon(x, y, z)^4$ are not constants. This is the case when the physical properties of the medium change from point to point, in particular, this happens in geometric optics when materials of different refractive indices are considered. In such case the Maxwell equations have the form:

$$\nabla \times \mathbf{E} = -\frac{\mu}{c} \frac{\partial \mathbf{H}}{\partial t}, \tag{5}$$

$$\nabla \times \mathbf{H} = \frac{2\pi}{c} \sigma \mathbf{E} + \frac{\epsilon}{c} \frac{\partial \mathbf{E}}{\partial t} \tag{6}$$

$$\nabla \cdot (\epsilon \mathbf{E}) = 4\pi\rho \tag{7}$$

$$\nabla \cdot (\mu \mathbf{H}) = 0, \tag{8}$$

c being the speed of light in vacuum. Recall that substances for which $\sigma \neq 0$ are conductors and if σ is negligibly small, the substances are called insulators or dielectrics, see [1, Sect. 1.1.2]. Under certain assumptions on the field and the physical set up we have that $\mathbf{J} = \sigma \mathbf{E}$, see [1, Sect. 1.1.2, formula (9)].

It is important to notice that these equations are written in Gaussian units, and the Maxwell equations in the first section written in SI units.

5.2   Maxwell Equations in Integral Form

Points in $\mathbb{R}^4$ are denoted by (x, y, z, t), and suppose $D \subset \mathbb{R}^4$ is a domain for which the divergence theorem holds, for example, the boundary is piecewise smooth, that is, a finite union of C^1 surfaces. For a point $P = (x, y, z, t)$ on the boundary ∂D, the unit outer normal at P is denoted by $\nu = (\nu_x, \nu_y, \nu_z, \nu_t)$. From equation (6)

$$\nabla \times \mathbf{H} - \frac{\epsilon}{c} \frac{\partial \mathbf{E}}{\partial t} = \frac{2\pi}{c} \sigma \mathbf{E}. \tag{9}$$

Recall we assume $\epsilon = \epsilon(x, y, z)$, and we want to derive an integral form of the last equation that does not require differentiability of the fields. In order to do that, we initially assume the fields are smooth and applying the divergence theorem we will obtain formulas independent of the derivatives of the fields. Set $\mathbf{H} = (\mathbf{H}_1, \mathbf{H}_2, \mathbf{H}_3)$. We have

---

[3]For values of μ for different substances see http://en.wikipedia.org/wiki/Permeability_ (electromagnetism)#Values_for_some_common_materials.

[4]For relative permittivity of some substances see http://en.wikipedia.org/wiki/Relative_ permittivity.

$$\int_D \nabla \times \mathbf{H} \, dx dy dz dt$$

$$= \mathbf{i} \int_D (\partial_y \mathbf{H}_3 - \partial_z \mathbf{H}_2) - \mathbf{j} \int_D (\partial_x \mathbf{H}_3 - \partial_z \mathbf{H}_1) + \mathbf{k} \int_D (\partial_x \mathbf{H}_2 - \partial_y \mathbf{H}_1)$$

$$= \mathbf{i} \int_D \text{div} \, (0, \mathbf{H}_3, -\mathbf{H}_2, 0) - \mathbf{j} \int_D \text{div} \, (\mathbf{H}_3, 0, -\mathbf{H}_1, 0) + \mathbf{k} \int_D \text{div} \, (\mathbf{H}_2, -\mathbf{H}_1, 0, 0)$$

$$= \mathbf{i} \int_{\partial D} (0, \mathbf{H}_3, -\mathbf{H}_2, 0) \cdot (v_x, v_y, v_z, v_t) \, d\sigma$$

$$- \mathbf{j} \int_{\partial D} (\mathbf{H}_3, 0, -\mathbf{H}_1, 0) \cdot (v_x, v_y, v_z, v_t) \, d\sigma + \mathbf{k} \int_{\partial D} (\mathbf{H}_2, -\mathbf{H}_1, 0, 0) \cdot (v_x, v_y, v_z, v_t) \, d\sigma$$

$$= \int_{\partial D} (v_x, v_y, v_z) \times (\mathbf{H}_1, \mathbf{H}_2, \mathbf{H}_3) \, d\sigma.$$

So integrating (9) over D yields

$$\int_{\partial D} (v_x, v_y, v_z) \times (\mathbf{H}_1, \mathbf{H}_2, \mathbf{H}_3) \, d\sigma - \int_D \frac{\epsilon}{c} \mathbf{E}_t \, dx dy dz dt$$

$$= \int_{\partial D} (v_x, v_y, v_z) \times (\mathbf{H}_1, \mathbf{H}_2, \mathbf{H}_3) \, d\sigma - \int_D \left(\frac{\epsilon}{c} \mathbf{E} \right)_t \, dx dy dz dt$$

$$= \int_{\partial D} (v_x, v_y, v_z) \times (\mathbf{H}_1, \mathbf{H}_2, \mathbf{H}_3) \, d\sigma - \int_{\partial D} \frac{\epsilon}{c} \mathbf{E} \, v_t \, d\sigma$$

$$= \int_{\partial D} \left((v_x, v_y, v_z) \times (\mathbf{H}_1, \mathbf{H}_2, \mathbf{H}_3) - \frac{\epsilon}{c} \mathbf{E} \, v_t \right) d\sigma = \int_D \frac{2\pi}{c} \sigma \mathbf{E} \, dx dy dz dt.$$

Therefore the surface integral

$$\int_{\partial D} \left((v_x, v_y, v_z) \times (\mathbf{H}_1, \mathbf{H}_2, \mathbf{H}_3) - \frac{\epsilon}{c} \mathbf{E} \, v_t \right) d\sigma = \int_D \frac{2\pi}{c} \sigma \mathbf{E} \, dx dy dz dt, \quad (10)$$

for each closed hyper-surface ∂D in $\mathbb{R}^4$. Proceeding in the same way with (5) we obtain that the surface integral

$$\int_{\partial D} \left((v_x, v_y, v_z) \times (\mathbf{E}_1, \mathbf{E}_2, \mathbf{E}_3) + \frac{\mu}{c} \mathbf{H} \, v_t \right) d\sigma = 0, \quad (11)$$

for each closed hyper-surface ∂D in $\mathbb{R}^4$.

Concerning (7) and (8), proceeding in the same way as before we obtain that

$$\int_{\partial D} \epsilon \mathbf{E} \cdot v \, d\sigma = 4\pi \int_D \rho \, dx dy dz dt \quad (12)$$

$$\int_{\partial D} \mu \mathbf{H} \cdot v \, d\sigma = 0, \quad (13)$$

for each domain $D \subset \mathbb{R}^4$ for which the divergence theorem holds. These formulas make sense as long as the fields $\mathbf{E}, \mathbf{H}$ and the coefficients μ and ϵ are piecewise continuous over ∂D and bounded. Equations (10)–(13) are Maxwell's equations in integral form.

5.3 Boundary Conditions at a Surface of Discontinuity

Let us consider a point $P_0 = (x_0, y_0, z_0, t_0)$, a hyper-surface Γ_0 passing through P_0 and suppose that the fields $\mathbf{H}$ and $\mathbf{E}$, solutions to the Maxwell equations in integral form, as well as the functions ϵ and μ, are discontinuous on Γ_0. Suppose that all these quantities are defined locally around P_0 say in the 4-dimensional ball $B_R(P_0)$. This situation is typical when we have two media with different indices of refraction and the surface Γ_0 is the one separating the two media. The surface Γ_0 divides the open ball $B_R(P_0)$ into two open pieces: B_R^+ and B_R^-. In order to make sense of the integrals we assume the surface Γ_0 is C^1, the fields $\mathbf{E}$ and $\mathbf{H}$, and ϵ and μ are bounded in $B_R(P_0)$, and all continuous on $B_R(P_0) \setminus \Gamma_0$. We assume also that for each $Q \in \Gamma_0 \cap B_R(P_0)$ the following limits exist and are finite:

$$\lim_{P \to Q, P \in B_R^+} \mathbf{E}(P) = \mathbf{E}^+(Q), \qquad \lim_{P \to Q, P \in B_R^+} \mathbf{H}(P) = \mathbf{H}^+(Q)$$

$$\lim_{P \to Q, P \in B_R^-} \mathbf{E}(P) = \mathbf{E}^-(Q), \qquad \lim_{P \to Q, P \in B_R^-} \mathbf{H}(P) = \mathbf{H}^-(Q),$$

and similar quantities for ϵ and μ. Let us call $\Gamma_+(R)$ the spherical part of the boundary of B_R^+, and $\Gamma_-(R)$ the spherical part of the boundary of B_R^-. If we let $\mathbf{E}^+(Q) = \mathbf{E}(Q)$ for $Q \in B_R^+$ and $\mathbf{E}^+(Q) = \lim_{P \to Q, P \in B_R^+} \mathbf{E}(P)$ for $Q \in \Gamma_0 \cap B_R$, and similarly $\mathbf{H}^+$, μ^+, and ϵ^+, then all these functions are continuous in the closure of B_R^+. In a similar way, we define $\mathbf{E}^-, \mathbf{H}^-, \mu_-$ and ϵ_- in B_R^- that are continuous in the closure of B_R^-. Hence by a limiting argument we can apply (10) with $D = B_R^\pm$ obtaining

$$\int_{\Gamma_+(R) \cup (\Gamma_0 \cap B_R(P_0))} \left((\nu_x, \nu_y, \nu_z) \times (\mathbf{H}_1^+, \mathbf{H}_2^+, \mathbf{H}_3^+) - \frac{\epsilon^+}{c} \mathbf{E}^+ \nu_t \right) d\sigma = \int_{B_R^+} \frac{2\pi}{c} \sigma \mathbf{E}^+ \, dx dy dz dt,$$

$$\tag{14}$$

and

$$\int_{\Gamma_-(R) \cup (\Gamma_0 \cap B_R(P_0))} \left((\nu_x, \nu_y, \nu_z) \times (\mathbf{H}_1^-, \mathbf{H}_2^-, \mathbf{H}_3^-) - \frac{\epsilon^-}{c} \mathbf{E}^- \nu_t \right) d\sigma = \int_{B_R^-} \frac{2\pi}{c} \sigma \mathbf{E}^- \, dx dy dz dt.$$

$$\tag{15}$$

Now

$$\int_{\Gamma_+(R)\cup(\Gamma_0\cap B_R(P_0))} \left((v_x, v_y, v_z) \times (\mathbf{H}_1^+, \mathbf{H}_2^+, \mathbf{H}_3^+) - \frac{\epsilon^+}{c}\mathbf{E}^+ \, v_t \right) d\sigma \quad (16)$$

$$= \int_{\Gamma_+(R)} \left((v_x, v_y, v_z) \times (\mathbf{H}_1, \mathbf{H}_2, \mathbf{H}_3) - \frac{\epsilon}{c}\mathbf{E} \, v_t \right) d\sigma$$

$$+ \int_{\Gamma_0\cap B_R(P_0)} \left((v_x, v_y, v_z) \times (\mathbf{H}_1^+, \mathbf{H}_2^+, \mathbf{H}_3^+) - \frac{\epsilon^+}{c}\mathbf{E}^+ \, v_t \right) d\sigma,$$

where in the integral over $\Gamma_0 \cap B_R(P_0)$, $v := (v_x, v_y, v_z, v_t)$ is the downward unit normal to Γ_0; and

$$\int_{\Gamma_-(R)\cup(\Gamma_0\cap B_R(P_0))} \left((v_x, v_y, v_z) \times (\mathbf{H}_1^-, \mathbf{H}_2^-, \mathbf{H}_3^-) - \frac{\epsilon^-}{c}\mathbf{E}^- \, v_t \right) d\sigma \quad (17)$$

$$= \int_{\Gamma_-(R)} \left((v_x, v_y, v_z) \times (\mathbf{H}_1, \mathbf{H}_2, \mathbf{H}_3) - \frac{\epsilon}{c}\mathbf{E} \, v_t \right) d\sigma$$

$$+ \int_{\Gamma_0\cap B_R(P_0)} \left((v_x, v_y, v_z) \times (\mathbf{H}_1^-, \mathbf{H}_2^-, \mathbf{H}_3^-) - \frac{\epsilon^-}{c}\mathbf{E}^- \, v_t \right) d\sigma,$$

where in the integral over $\Gamma_0 \cap B_R(P_0)$, v is the upward unit normal to Γ_0. Adding (16) and (17), and using (14) and (15) yields

$$\int_{B_R^+} \frac{2\pi}{c}\sigma\mathbf{E}^+ \, dxdydzdt + \int_{B_R^-} \frac{2\pi}{c}\sigma\mathbf{E}^- \, dxdydzdt$$

$$= \int_{\Gamma_+(R)} \left((v_x, v_y, v_z) \times (\mathbf{H}_1, \mathbf{H}_2, \mathbf{H}_3) - \frac{\epsilon}{c}\mathbf{E} \, v_t \right) d\sigma + \int_{\Gamma_-(R)} \left((v_x, v_y, v_z) \times (\mathbf{H}_1, \mathbf{H}_2, \mathbf{H}_3) - \frac{\epsilon}{c}\mathbf{E} \, v_t \right) d\sigma$$

$$+ \int_{\Gamma_0\cap B_R(P_0)} \left((v_x, v_y, v_z) \times (\mathbf{H}^+ - \mathbf{H}^-) - \frac{1}{c}\left(\epsilon^+ \mathbf{E}^+ - \epsilon^- \mathbf{E}^-\right) v_t \right) d\sigma.$$

On the other hand, applying (10) with $D = B_R$ yields

$$\int_{\Gamma_+(R)\cup\Gamma_-(R)} \left((v_x, v_y, v_z) \times (\mathbf{H}_1, \mathbf{H}_2, \mathbf{H}_3) - \frac{\epsilon}{c}\mathbf{E} \, v_t \right) d\sigma = \int_{B_R} \frac{2\pi}{c}\sigma\mathbf{E} \, dxdydzdt.$$

Since the field $\mathbf{E}$ is discontinuous only on Γ_0, which is a set of measure zero, we therefore obtain

$$\int_{\Gamma_0\cap B_R(P_0)} \left((v_x, v_y, v_z) \times (\mathbf{H}^+ - \mathbf{H}^-) - \frac{1}{c}\left(\epsilon^+ \mathbf{E}^+ - \epsilon^- \mathbf{E}^-\right) v_t \right) d\sigma = 0$$

for all R sufficiently small. Now letting $R \to 0$ we obtain the following equation valid at P_0

$$(v_x, v_y, v_z) \times (\mathbf{H}^+ - \mathbf{H}^-) - \frac{1}{c}(\epsilon^+ \mathbf{E}^+ - \epsilon^- \mathbf{E}^-) v_t = 0, \qquad (18)$$

where v is the normal to the interface Γ_0 at the point P_0.

Suppose the interface is independent of time and is given by a function $\phi(x, y, z) = 0$, then the normal at a point is $v = (\phi_x, \phi_y, \phi_z, 0)$, therefore (18) becomes

$$\nabla\phi \times (\mathbf{H}^+ - \mathbf{H}^-) = 0.$$

We can write $\mathbf{H}^\pm = \mathbf{H}^\pm_{\tan} + \mathbf{H}^\pm_{\text{perp}}$, where $\mathbf{H}^\pm_{\text{perp}}$ is the component in the direction of the normal $\nabla\phi$, and $\mathbf{H}^\pm_{\tan}$ is the component perpendicular to the normal. We have $\nabla\phi \times \mathbf{H}^\pm = \nabla\phi \times \mathbf{H}^\pm_{\tan} + \nabla\phi \times \mathbf{H}^\pm_{\text{perp}} = \nabla\phi \times \mathbf{H}^\pm_{\tan}$. So

$$0 = \nabla\phi \times (\mathbf{H}^+ - \mathbf{H}^-) = \nabla\phi \times (\mathbf{H}^+_{\tan} - \mathbf{H}^-_{\tan}) = |\nabla\phi| \, |\mathbf{H}^+_{\tan} - \mathbf{H}^-_{\tan}|,$$

since the vectors are perpendicular. So if $\nabla\phi \neq 0$, then we obtain the important relation that

$$\mathbf{H}^+_{\tan} - \mathbf{H}^-_{\tan} = 0,$$

that is, *the tangential components of the magnetic field are continuous across the boundary.*

Proceeding in the same manner this time with (11) yields the equation

$$(v_x, v_y, v_z) \times (\mathbf{E}^+ - \mathbf{E}^-) + \frac{1}{c}(\mu^+ \mathbf{H}^+ - \mu^- \mathbf{H}^-) v_t = 0, \qquad (19)$$

where v is the normal to the interface Γ_0 at the point P_0. If the interface is independent of t, proceeding exactly as before, we obtain

$$\nabla\phi \times (\mathbf{E}^+ - \mathbf{E}^-) = 0,$$

and

$$\mathbf{E}^+_{\tan} - \mathbf{E}^-_{\tan} = 0,$$

that is, also *the tangential components of the electric field are continuous across the boundary.*

In regard to (12) and (13), we obtain similarly that

$$(\epsilon^+ \mathbf{E}^+ - \epsilon^- \mathbf{E}^-) \cdot v = 0, \text{ and } (\mu^+ \mathbf{H}^+ - \mu^- \mathbf{H}^-) \cdot v = 0.$$

Since $\mathbf{H}^\pm \cdot v = \mathbf{H}^\pm_{\text{perp}} \cdot v$, and similarly for $\mathbf{E}$, assuming $\phi = \phi(x, y, z)$ yields

$$0 = (\epsilon^+ \mathbf{E}_{\text{perp}}^+ - \epsilon^- \mathbf{E}_{\text{perp}}^-) \cdot \nabla\phi = |\epsilon^+ \mathbf{E}_{\text{perp}}^+ - \epsilon^- \mathbf{E}_{\text{perp}}^-| \, |\nabla\phi|,$$

and

$$0 = (\mu^+ \mathbf{H}_{\text{perp}}^+ - \mu^- \mathbf{H}_{\text{perp}}^-) \cdot \nabla\phi = |\mu^+ \mathbf{H}_{\text{perp}}^+ - \mu^- \mathbf{H}_{\text{perp}}^-| \, |\nabla\phi|,$$

and therefore

$$|\epsilon^+ \mathbf{E}_{\text{perp}}^+ - \epsilon^- \mathbf{E}_{\text{perp}}^-| = |\mu^+ \mathbf{H}_{\text{perp}}^+ - \mu^- \mathbf{H}_{\text{perp}}^-| = 0.$$

Therefore *the perpendicular components of the fields $\epsilon\mathbf{E}$ and $\mu\mathbf{H}$ are continuous across the interface.*

5.4   Maxwell's Equations in the Absence of Charges

This is the case when $\rho = 0$ and $\mathbf{J} = 0$. So the equations become

$$\nabla \cdot \mathbf{E} = 0, \tag{20}$$

$$\nabla \cdot \mathbf{H} = 0, \tag{21}$$

$$\nabla \times \mathbf{E} = -\frac{\partial \mathbf{H}}{\partial t}, \tag{22}$$

$$\nabla \times \mathbf{H} = \epsilon_0 \mu_0 \frac{\partial \mathbf{E}}{\partial t}, \tag{23}$$

5.5   The Wave Equation

Recall the following formula from vector analysis for a vector $\mathbf{A} = \mathbf{A}(x, y, z) = (\mathbf{A}_x(x, y, z), \mathbf{A}_y(x, y, z), \mathbf{A}_z(x, y, z))$:

$$\nabla \times (\nabla \times \mathbf{A}) = \nabla(\nabla \cdot \mathbf{A}) - (\nabla \cdot \nabla)\mathbf{A}. \tag{24}$$

Denote $\nabla \cdot \nabla = \nabla^2$, the Laplacian, and so

$$\nabla^2 \mathbf{A} =$$

$$\left(\frac{\partial^2 \mathbf{A}_x}{\partial x^2} + \frac{\partial^2 \mathbf{A}_x}{\partial y^2} + \frac{\partial^2 \mathbf{A}_x}{\partial z^2} \right) \mathbf{i} + \left(\frac{\partial^2 \mathbf{A}_y}{\partial x^2} + \frac{\partial^2 \mathbf{A}_y}{\partial y^2} + \frac{\partial^2 \mathbf{A}_y}{\partial z^2} \right) \mathbf{j} + \left(\frac{\partial^2 \mathbf{A}_z}{\partial x^2} + \frac{\partial^2 \mathbf{A}_z}{\partial y^2} + \frac{\partial^2 \mathbf{A}_z}{\partial z^2} \right) \mathbf{k}.$$

From Faraday's law and Ampère's law

$$\nabla \times (\nabla \times \mathbf{E}) = -\frac{\partial(\nabla \times \mathbf{H})}{\partial t} = -\epsilon_0 \mu_0 \frac{\partial^2 \mathbf{E}}{\partial t^2}$$

and so from formula (24) we obtain that $\mathbf{E}$ satisfies the wave equation

$$\epsilon_0 \mu_0 \frac{\partial^2 \mathbf{E}}{\partial t^2} = \nabla^2 \mathbf{E}.$$

Proceeding in the same manner for $\mathbf{H}$ we obtain

$$\epsilon_0 \mu_0 \frac{\partial^2 \mathbf{H}}{\partial t^2} = \nabla^2 \mathbf{H}.$$

That is, both the electric and magnetic fields satisfy the wave equation. We have from physical considerations that

$$v = c = \frac{1}{\sqrt{\epsilon_0 \mu_0}},$$

c being the speed of propagation of light in free space, which in this case is the velocity v of propagation. If free space is changed by a material with other values of μ_0 and ϵ_0, the velocity v represent the speed of propagation of waves in this material.[5]

5.6 Dispersion Equation

Suppose $\mathbf{E}$ and $\mathbf{H}$ solve the Maxwell equations (5)–(8) with $\rho = 0$, $\sigma = 0$, and ϵ and μ constants. Then

$$\nabla \times (\nabla \times \mathbf{E}) = -\frac{\mu}{c} \nabla \times \mathbf{H}_t = -\frac{\mu}{c} (\nabla \times \mathbf{H})_t = -\frac{\epsilon\mu}{c^2} \mathbf{E}_{tt}.$$

On the other hand, from (24), $\nabla \times (\nabla \times \mathbf{E}) = -\nabla^2 \mathbf{E}$ and so

$$\nabla^2 \mathbf{E} - \frac{\epsilon\mu}{c^2} \mathbf{E}_{tt} = 0,$$

and similarly

$$\nabla^2 \mathbf{H} - \frac{\epsilon\mu}{c^2} \mathbf{H}_{tt} = 0.$$

---

[5]The relative permittivity is ϵ/ϵ_0 and the relative permeability is μ/μ_0; the index of refraction is defined by $n = \sqrt{\epsilon_r \mu_r}$. The velocity of propagation $v = \frac{1}{\sqrt{\epsilon\mu}}$. Since $c = \frac{1}{\sqrt{\epsilon_0 \mu_0}}$, we get that $n = c/v$.

If $\mathbf{E} = A \cos{(\mathbf{r} \cdot \mathbf{k} + \omega t)}$, then we obtain the dispersion equation

$$|\mathbf{k}|^2 = \epsilon\mu \left(\frac{\omega}{c}\right)^2 . \tag{25}$$

5.7 Plane Waves

Let $\mathbf{s}$ be a unit vector. Any solution to the wave equation

$$\frac{1}{v^2}\partial_t^2 V = \nabla^2 V,$$

of the form $V(\mathbf{r}, t) = F(\mathbf{r} \cdot \mathbf{s}, t)$ is called a *plane wave*, since at each time t, V is constant on each plane of the form $\mathbf{r} \cdot \mathbf{s} =$ constant. That is, for each t the vector $V(\mathbf{r}, t)$ is the same on each plane $\mathbf{r} \cdot \mathbf{s} =$ constant. The plane wave propagates in the direction $\mathbf{s}$. It can be proved that any solution to the wave equation of this form can be written as

$$V(\mathbf{r}, t) = V_1(\mathbf{r} \cdot \mathbf{s} - vt) + V_2(\mathbf{r} \cdot \mathbf{s} + vt)$$

where V_1, V_2 are arbitrary functions, see [1, Sect. 1.3.1]. Since the fields $\mathbf{E}$ and $\mathbf{H}$ both satisfy the wave equation, it is then natural to consider the case when

$$\mathbf{E} = \mathbf{E}(\mathbf{r} \cdot \mathbf{s} - vt), \quad \mathbf{H} = \mathbf{H}(\mathbf{r} \cdot \mathbf{s} - vt),$$

that is, $\mathbf{E}$ and $\mathbf{H}$ are functions of the scalar variable $\mathbf{r} \cdot \mathbf{s} - vt$. We have

$$\frac{\partial \mathbf{E}}{\partial t} = -v\mathbf{E}', \text{ and } \nabla \times \mathbf{E} = s \times \mathbf{E}';$$

and similarly for $\mathbf{H}$ under the assumption that $\mathbf{J} = 0$. Thus, from the Faraday and Ampère laws, and since $v^2 = \dfrac{1}{\epsilon_0\mu_0}$, we obtain the equations

$$s \times \mathbf{E}' = v\mathbf{H}'$$

$$s \times \mathbf{H}' = -\frac{1}{v}\mathbf{E}'.$$

Since s is a constant vector $s \times \mathbf{E}' = (s \times \mathbf{E})'$, and so the equations are

$$(s \times \mathbf{E})' = v\mathbf{H}'$$

$$(s \times \mathbf{H})' = -\frac{1}{v}\mathbf{E}'.$$

Integrating these equations and taking constants of integration zero (which amounts to neglect constant fields), we obtain the very important equations relating the electric and magnetic fields

$$\mathbf{E} = -v(s \times \mathbf{H}) \tag{26}$$

$$\mathbf{H} = \frac{1}{v}(s \times \mathbf{E}). \tag{27}$$

This shows that $\mathbf{s} \cdot \mathbf{E} = \mathbf{s} \cdot \mathbf{H} = 0$, that means, the electric and magnetic field are always *perpendicular* to the direction of propagation $\mathbf{s}$. In addition, $\mathbf{E} \cdot \mathbf{H} = v(s \times \mathbf{H}) \cdot \mathbf{H} = 0$, that is, $\mathbf{E}$ and $\mathbf{H}$ are always perpendicular. We also obtain taking absolute values that

$$|\mathbf{E}| = v|\mathbf{H}|.$$

5.8 Fresnel Formulas

We consider plane waves whose components have the form

$$a \cos \left(\omega \left(t - \frac{\mathbf{r} \cdot \mathbf{s}}{v} \right) + \delta \right) = a \cos \left(\omega t - \mathbf{k} \cdot \mathbf{r} + \delta \right),$$

that is, $\mathbf{k} = \dfrac{\omega}{v}\mathbf{s}$ and a, δ are real numbers. The quantity $\omega t - \mathbf{k} \cdot \mathbf{r} + \delta$ is called the *phase*, and a is called the *amplitude*.

Let $\mathbf{s}^i$ be the direction (unit) of an incident plane wave traveling for a while in medium I with velocity of propagation v_1 that hits, at a point P, a boundary Γ between I and another medium II where the velocity of propagation is v_2 (I and II are also called dielectrics as they are materials with zero conductivity, that is $\sigma = 0$ and so the current density vector $\mathbf{J} = 0$, see Sect. 5.1). Then the wave splits into two waves: a *transmitted wave* propagating in medium II and a *reflected wave* propagated back into medium I. We shall assume that these two waves are also plane. The plane determined by v and $\mathbf{s}^i$ is called the *incidence plane*.

It is important to remark that for our analysis, we will choose a local system of coordinates around the point P. Indeed, we are going to write all fields as functions of $\mathbf{r}$ (position) and t, with $\mathbf{r}$ close to zero, such that the coordinates of P in this system are $\mathbf{r} = 0$. In particular, the fields will be calculated near the point P.

We choose a rectangular right-hand system of coordinates x, y, z such that the normal v is on the z-axis, and the x and y axes are on the plane perpendicular to v and in such a way that the vector $\mathbf{s}^i$ lies on the xz-plane. This means that the tangent plane to Γ at P is the xy-plane; in particular, $P = (0, 0, 0)$. So we assume that

$$\mathbf{s}^i = \sin \theta_i \mathbf{i} + \cos \theta_i \mathbf{k}$$

that is, $\mathbf{s}^i$ lives on the xz-plane and so the direction of propagation is perpendicular to the y-axis and θ_i is the angle between the normal vector ν to the boundary at P (the z-axis) and the incident direction $\mathbf{s}^i$ (as usual $\mathbf{i}, \mathbf{j}, \mathbf{k}$ are the unit coordinate vectors).

The electric field corresponding to this incident field is

$$\mathbf{E}^i(\mathbf{r}, t) = \left(-I_\parallel \cos\theta_i, I_\perp, I_\parallel \sin\theta_i\right) \cos\left(\omega\left(t - \frac{\mathbf{r}\cdot\mathbf{s}^i}{v_1}\right)\right) = \mathbf{E}_0^i \cos\left(\omega\left(t - \frac{\mathbf{r}\cdot\mathbf{s}^i}{v_1}\right)\right). \tag{28}$$

Notice that $\mathbf{E}$ has this form because, as is was proved in Sect. 5.7, $\mathbf{E}$ is always perpendicular to the direction of propagation $\mathbf{s}^i$. Notice also that the field $\mathbf{E}^i$ has a component that is perpendicular to the plane of incidence and a component that is parallel to this plane, indeed, we write

$$\mathbf{E}_\perp^i = I_\perp \cos\left(\omega\left(t - \frac{\mathbf{r}\cdot\mathbf{s}^i}{v_1}\right)\right)\mathbf{j},$$

and

$$\mathbf{E}_\parallel^i = \left(-I_\parallel \cos\theta_i\,\mathbf{i} + I_\parallel \sin\theta_i\,\mathbf{k}\right)\cos\left(\omega\left(t - \frac{\mathbf{r}\cdot\mathbf{s}^i}{v_1}\right)\right).$$

Also notice that

$$|\mathbf{E}^i|^2 = \left(I_\parallel^2 + I_\perp^2\right)\cos^2\left(\omega\left(t - \frac{\mathbf{r}\cdot\mathbf{s}^i}{v_1}\right)\right)$$

From (27), the magnetic field is then

$$\mathbf{H}^i(\mathbf{r}, t) = \frac{1}{v_1}\left(-I_\perp \cos\theta_i, -I_\parallel, I_\perp \sin\theta_i\right)\cos\left(\omega\left(t - \frac{\mathbf{r}\cdot\mathbf{s}^i}{v_1}\right)\right) = \mathbf{H}_0^i \cos\left(\omega\left(t - \frac{\mathbf{r}\cdot\mathbf{s}^i}{v_1}\right)\right).$$

Let us now introduce $\mathbf{s}^t$, the direction of propagation of the transmitted wave, and θ_t the angle between the normal ν and $\mathbf{s}^t$, and similarly, $\mathbf{s}^r$ is the direction of propagation of the reflected wave and θ_r is the angle between the normal ν and $\mathbf{s}^r$. We have from the Snell law that $\mathbf{s}^r = \sin\theta_r\,\mathbf{i} + \cos\theta_r\,\mathbf{k} = \sin\theta_i\,\mathbf{i} - \cos\theta_i\,\mathbf{k}$. Then the electric and magnetic fields corresponding to transmission are

$$\mathbf{E}^t(\mathbf{r}, t) = \left(-T_\parallel \cos\theta_t, T_\perp, T_\parallel \sin\theta_t\right)\cos\left(\omega\left(t - \frac{\mathbf{r}\cdot\mathbf{s}^t}{v_2}\right)\right) = \mathbf{E}_0^t \cos\left(\omega\left(t - \frac{\mathbf{r}\cdot\mathbf{s}^t}{v_2}\right)\right) \tag{29}$$

$$\mathbf{H}^t(\mathbf{r}, t) = \frac{1}{v_2}\left(-T_\perp \cos\theta_t, -T_\parallel, T_\perp \sin\theta_t\right)\cos\left(\omega\left(t - \frac{\mathbf{r}\cdot\mathbf{s}^t}{v_2}\right)\right) = \mathbf{H}_0^t \cos\left(\omega\left(t - \frac{\mathbf{r}\cdot\mathbf{s}^t}{v_2}\right)\right);$$

and similarly the fields corresponding to reflection are

$$\mathbf{E}^r(\mathbf{r}, t) = (-R_\parallel \cos \theta_r, R_\perp, R_\parallel \sin \theta_r) \cos \left(\omega \left(t - \frac{\mathbf{r} \cdot \mathbf{s}^r}{v_1} \right) \right) = \mathbf{E}_0^r \cos \left(\omega \left(t - \frac{\mathbf{r} \cdot \mathbf{s}^r}{v_1} \right) \right) \quad (30)$$

$$\mathbf{H}^r(\mathbf{r}, t) = \frac{1}{v_1} (-R_\perp \cos \theta_r, -R_\parallel, R_\perp \sin \theta_r) \cos \left(\omega \left(t - \frac{\mathbf{r} \cdot \mathbf{s}^r}{v_1} \right) \right) = \mathbf{H}_0^r \cos \left(\omega \left(t - \frac{\mathbf{r} \cdot \mathbf{s}^r}{v_1} \right) \right).$$

Recall that from Snell's law all vectors $\mathbf{s}^i, \mathbf{s}^t, \mathbf{s}^r$ and v all live on the same plane, that is, the xz-plane. Each of the electric and magnetic fields can be decomposed uniquely as a sum of a component in the direction of the normal (normal component) or on the z-axis, plus another component perpendicular to the normal (tangential component) or on the xy-plane. From the integral form of Maxwell's equations, as it shown in Sect. 5.3, the tangential components of $\mathbf{E}$ (and also of $\mathbf{H}$ if $\mathbf{J} = 0$) at the interface are continuous (see also [1, Sect. 1.1.3, formula (23)]). Since the electric field on medium I near Γ equals $\mathbf{E}^i + \mathbf{E}^r$, we get $\mathbf{E}_{\text{tan}}^i + \mathbf{E}_{\text{tan}}^r = \mathbf{E}_{\text{tan}}^t$ on Γ, since $\mathbf{E}_{\text{tan}}^t$ is the transmitted electric field in medium II near Γ. From the configuration we have, we can write $\mathbf{E}^i = \mathbf{E}_{\text{normal}}^i \mathbf{k} + \mathbf{E}_{\text{tan}}^i$, and so $\mathbf{k} \times \mathbf{E}^i = \mathbf{k} \times \mathbf{E}_{\text{tan}}^i$. Similarly, $\mathbf{k} \times \mathbf{E}^r = \mathbf{k} \times \mathbf{E}_{\text{tan}}^r$ and $\mathbf{k} \times \mathbf{E}^t = \mathbf{k} \times \mathbf{E}_{\text{tan}}^t$. So $\mathbf{k} \times \mathbf{E}^i + \mathbf{k} \times \mathbf{E}^r = \mathbf{k} \times \mathbf{E}^t$. Then

$$\mathbf{k} \times \mathbf{E}_0^i \cos \left(\omega \left(t - \frac{\mathbf{r} \cdot \mathbf{s}^i}{v_1} \right) \right) + \mathbf{k} \times \mathbf{E}_0^r \cos \left(\omega \left(t - \frac{\mathbf{r} \cdot \mathbf{s}^r}{v_1} \right) \right) = \mathbf{k} \times \mathbf{E}_0^t \cos \left(\omega \left(t - \frac{\mathbf{r} \cdot \mathbf{s}^t}{v_2} \right) \right),$$

for all $\mathbf{r}$ close to zero and all t. The interface point P is $\mathbf{r} = (0, 0, 0)$, so in particular, we obtain

$$\mathbf{k} \times \mathbf{E}_0^i \cos(\omega t) + \mathbf{k} \times \mathbf{E}_0^r \cos(\omega t) = \mathbf{k} \times \mathbf{E}_0^t \cos(\omega t)$$

for all t. Eliminating the cosines we get

$$\mathbf{k} \times \mathbf{E}_0^i + \mathbf{k} \times \mathbf{E}_0^r = \mathbf{k} \times \mathbf{E}_0^t. \quad (31)$$

Since we are assuming the current density vector $\mathbf{J} = 0$, as it was mentioned earlier, the tangential component of the magnetic field is also continuous across the interface. So as before with the electric field, we have $\mathbf{H}_{\text{tan}}^i + \mathbf{H}_{\text{tan}}^r = \mathbf{H}_{\text{tan}}^t$ on Γ, and so

$$\mathbf{k} \times \mathbf{H}_0^i + \mathbf{k} \times \mathbf{H}_0^r = \mathbf{k} \times \mathbf{H}_0^t. \quad (32)$$

From (31) we obtain the equations

$$I_\perp + R_\perp = T_\perp, \qquad \cos \theta_i (I_\parallel - R_\parallel) = \cos \theta_t T_\parallel;$$

and from (32) we obtain

$$\frac{I_\parallel}{v_1} + \frac{R_\parallel}{v_1} = \frac{T_\parallel}{v_2}, \qquad \cos \theta_i \left(\frac{I_\perp}{v_1} - \frac{R_\perp}{v_1} \right) = \cos \theta_t \frac{T_\perp}{v_2}.$$

We have $n_1 = c/v_1$ and $n_2 = c/v_2$ so solving the last two sets of equations yields

$$T_\| = \frac{2n_1 \cos\theta_i}{n_2 \cos\theta_i + n_1 \cos\theta_t} I_\|$$

$$T_\perp = \frac{2n_1 \cos\theta_i}{n_1 \cos\theta_i + n_2 \cos\theta_t} I_\perp$$

$$R_\| = \frac{n_2 \cos\theta_i - n_1 \cos\theta_t}{n_2 \cos\theta_i + n_1 \cos\theta_t} I_\|$$

$$R_\perp = \frac{n_1 \cos\theta_i - n_2 \cos\theta_t}{n_1 \cos\theta_i + n_2 \cos\theta_t} I_\perp.$$

These are the *Fresnel equations* expressing the amplitudes of the reflected and transmitted waves in terms of the amplitude of the incident wave.

5.9 Rewriting the Fresnel Equations

We will replace $\mathbf{s^i}$ by x and $\mathbf{s^t}$ by m, and we also set $\kappa = n_2/n_1$. Recall ν is the normal to the interface. We have $\cos\theta_i = x \cdot \nu$ and $\cos\theta_t = m \cdot \nu$. In addition, from the Snell law $x - \kappa m = \lambda\nu$, so the Fresnel equations have the form

$$T_\| = \frac{2x \cdot \nu}{\kappa x \cdot \nu + m \cdot \nu} I_\| = \frac{2x \cdot \nu}{(\kappa x + m) \cdot \nu} I_\| = \frac{2x \cdot (x - \kappa m)}{(\kappa x + m) \cdot (x - \kappa m)} I_\|$$

$$T_\perp = \frac{2x \cdot \nu}{x \cdot \nu + \kappa m \cdot \nu} I_\perp = \frac{2x \cdot \nu}{(x + \kappa m) \cdot \nu} I_\perp = \frac{2x \cdot (x - \kappa m)}{(x + \kappa m) \cdot (x - \kappa m)} I_\perp$$

$$R_\| = \frac{\kappa x \cdot \nu - m \cdot \nu}{\kappa x \cdot \nu + m \cdot \nu} I_\| = \frac{(\kappa x - m) \cdot \nu}{(\kappa x + m) \cdot \nu} I_\| = \frac{(\kappa x - m) \cdot (x - \kappa m)}{(\kappa x + m) \cdot (x - \kappa m)} I_\|$$

$$R_\perp = \frac{x \cdot \nu - \kappa m \cdot \nu}{x \cdot \nu + \kappa m \cdot \nu} I_\perp = \frac{(x - \kappa m) \cdot \nu}{(x + \kappa m) \cdot \nu} I_\perp = \frac{(x - \kappa m) \cdot (x - \kappa m)}{(x + \kappa m) \cdot (x - \kappa m)} I_\perp.$$

Notice that the denominators of the perpendicular components are the same and likewise for the parallel components.

5.10 The Poynting Vector

It is defined by

$$\mathbf{S} = \frac{c}{4\pi} \mathbf{E} \times \mathbf{H},$$

where c is the speed of light in free space. The vector $\mathbf{S}$ represents the flux of energy through a surface. Suppose dA is the area of a surface element at a point P and let

ν be the normal at P. Then the flux of energy through dA at the point P is given by

$$dF = \mathbf{S} \cdot \nu \, dA.$$

From (27) we get that

$$\mathbf{S} = \frac{c}{4\pi v} \mathbf{E} \times (\mathbf{s} \times \mathbf{E}) = \frac{n}{4\pi} |\mathbf{E}|^2 \mathbf{s}.$$

Using the form of the incident wave from the previous section, the amount of energy J^i flowing through a unit area of the boundary per second at P, of the incident wave $\mathbf{E}^i$ given in (28), is then

$$J^i = |\mathbf{S}^i| \cos \theta_i = \frac{n_1}{4\pi} |\mathbf{E}_0^i|^2 \cos \theta_i.^6$$

Similarly, the amount of energy in the reflected and transmitted waves (also given in the previous section) leaving a unit area of the boundary per second at P are given by

$$J^r = |\mathbf{S}^r| \cos \theta_i = \frac{n_1}{4\pi} |\mathbf{E}_0^r|^2 \cos \theta_i$$

$$J^t = |\mathbf{S}^t| \cos \theta_t = \frac{n_2}{4\pi} |\mathbf{E}_0^t|^2 \cos \theta_t.$$

The reflection and transmission coefficients are defined by

$$\mathcal{R} = \frac{J^r}{J^i} = \left(\frac{|\mathbf{E}_0^r|}{|\mathbf{E}_0^i|} \right)^2, \text{ and } \mathcal{T} = \frac{J^t}{J^i} = \frac{n_2 \cos \theta_t}{n_1 \cos \theta_i} \left(\frac{|\mathbf{E}_0^t|}{|\mathbf{E}_0^i|} \right)^2. \tag{33}$$

By conservation of energy or by direct verification $\mathcal{R} + \mathcal{T} = 1$.

5.11 Polarization

Polarization is a property of the field that describes the orientation of their oscillations. Since the electric vector is assumed a plane wave and as we showed it is perpendicular to the direction of propagation $\mathbf{s}$, then for each $\mathbf{r}$ in the plane

---

[6]Notice that from (28), the value of the field $\mathbf{E}^i$ at P is $\mathbf{E}^i(0, t) = \mathbf{E}_0^i \cos(\omega t)$. Hence we actually get $J^i = \frac{n_1}{4\pi} |\mathbf{E}_0^i|^2 \cos \theta_i \, \cos^2(\omega t)$. Similarly, from (29), the value of the field $\mathbf{E}^t$ at P is $\mathbf{E}^t(0, t) = \mathbf{E}_0^t \cos(\omega t)$, and from (30), the value of the field $\mathbf{E}^r$ at P is $\mathbf{E}^r(0, t) = \mathbf{E}_0^r \cos(\omega t)$. Therefore we have $J^r = \frac{n_1}{4\pi} |\mathbf{E}_0^r|^2 \cos \theta_i \, \cos^2(\omega t)$, and $J^t = \frac{n_2}{4\pi} |\mathbf{E}_0^t|^2 \cos \theta_t \, \cos^2(\omega t)$. So we get formulas (33) because the factor $\cos^2(\omega t)$ cancels out.

$\mathbf{r} \cdot \mathbf{s} = c$ and t fixed, the vector $\mathbf{E}(\mathbf{r}, t)$ is constant. We visualize $\mathbf{E}(\mathbf{r}, t)$ as a vector with origin at the intersection of the direction $\mathbf{s}$ with the plane $\mathbf{r} \cdot \mathbf{s} = c$. That is, as a vector with origin at the point $(\mathbf{r} \cdot \mathbf{s})\mathbf{s}$ and terminal point $(\mathbf{r} \cdot \mathbf{s})\mathbf{s} + \mathbf{E}(\mathbf{r}, t)$. Then t is fixed and $\mathbf{r}$ runs over all space, the end point of this vector describes a curve in 3-d. If we now move t, this curve is shifted (and keeps the same shape) by changing the phase because of the presence of ωt in the cos function. So when $\mathbf{r} \cdot \mathbf{s} = c$ and t moves the vector $\mathbf{E}(\mathbf{r}, t)$ describes a curve in the plane $\mathbf{r} \cdot \mathbf{s} = c$. If this curve is an ellipse we say that the wave is *elliptically polarized* and when the ellipse is a circle we say the wave is *circularly polarized*, and if the ellipse degenerates to a segment we say the wave is *linearly polarized*. If the wave describing the incident field has components that have different phases, then this changes the sense of circulation and inclination of the ellipse (for elliptically polarized light), see [1, Sect. 1.4.2]. See http://en.wikipedia.org/wiki/Polarization for pictures.

Suppose for example that the wave is linearly polarized perpendicularly to the plane of incidence. That is, $I_\parallel = 0$. Then from Fresnel equations $T_\parallel = R_\parallel = 0$ and

$$\mathcal{R} = \left(\frac{|\mathbf{E}_0^r|}{|\mathbf{E}_0^i|} \right)^2 = \left(\frac{R_\perp}{I_\perp} \right)^2 = \left(\frac{|x - \kappa m|^2}{1 - \kappa^2} \right)^2,$$

and

$$\mathcal{T} = \kappa \frac{m \cdot v}{x \cdot v} \left(\frac{T_\perp}{I_\perp} \right)^2 = \kappa \frac{m \cdot (x - \kappa m)}{x \cdot (x - \kappa m)} \left(\frac{2 x \cdot (x - \kappa m)}{(x + \kappa m) \cdot (x - \kappa m)} \right)^2$$

$$= \frac{4\kappa}{(1 - \kappa^2)^2} \left(m \cdot (x - \kappa m) \right) (x \cdot (x - \kappa m)).$$

For the case when no polarization is assumed, that is, radiation has no particular preference for the direction in which it vibrates, we have from Fresnel's equations that

$$|\mathbf{E}_0^r|^2 = R_\parallel^2 + R_\perp^2 = \left[\frac{(\kappa x - m) \cdot (x - \kappa m)}{(\kappa x + m) \cdot (x - \kappa m)} \right]^2 I_\parallel^2 + \left[\frac{(x - \kappa m) \cdot (x - \kappa m)}{(x + \kappa m) \cdot (x - \kappa m)} \right]^2 I_\perp^2,$$

and so

$$\mathcal{R} = \left(\frac{|\mathbf{E}_0^r|}{|\mathbf{E}_0^i|} \right)^2 = \frac{R_\parallel^2 + R_\perp^2}{I_\parallel^2 + I_\perp^2}$$

$$= \left[\frac{(\kappa x - m) \cdot (x - \kappa m)}{(\kappa x + m) \cdot (x - \kappa m)} \right]^2 \frac{I_\parallel^2}{I_\parallel^2 + I_\perp^2} + \left[\frac{(x - \kappa m) \cdot (x - \kappa m)}{(x + \kappa m) \cdot (x - \kappa m)} \right]^2 \frac{I_\perp^2}{I_\parallel^2 + I_\perp^2}$$

$$= \frac{1}{(1 - \kappa^2)^2} \left(\left[\frac{2\kappa}{x \cdot m} - (1 + \kappa^2) \right]^2 \frac{I_\parallel^2}{I_\parallel^2 + I_\perp^2} + \left[1 - 2\kappa x \cdot m + \kappa^2 \right]^2 \frac{I_\perp^2}{I_\parallel^2 + I_\perp^2} \right)$$

which is a function only of $x \cdot m$. In principle the coefficients $I_{\parallel}$ and $I_{\perp}$ might depend on the direction x, in other words, for each direction x we would have a wave that changes its amplitude with the direction of propagation. The energy of the incident wave would be $f(x) = |\mathbf{E}_0^i|^2 = I_{\parallel}(x)^2 + I_{\perp}(x)^2$. Notice that if the incidence is normal, that is, $x = m$, then $\mathcal{R} = \left(\dfrac{1-\kappa}{1+\kappa}\right)^2$ which shows that even for radiation normal to the interface we lose energy by reflection. For example, if we go from air to glass, $n_1 = 1$ and $n_2 = 1.5$, we have $\kappa = 1.5$ so $\mathcal{R} = .04$, which means that 4% of the energy is lost in internal reflection.

5.12   Estimation of the Fresnel Coefficients

For later purposes we need to estimate the function

$$\phi(s) = \frac{1}{(1-\kappa^2)^2} \left(\left[\frac{2\kappa}{s} - (1+\kappa^2) \right]^2 \alpha + \left[1 - 2\kappa s + \kappa^2 \right]^2 \beta \right), \qquad (34)$$

have $0 \leq \alpha, \beta \leq 1$ and $\alpha + \beta = 1$. Set

$$g(t) = \left[\frac{2\kappa}{t} - (1+\kappa^2) \right]^2, \qquad h(t) = \left[1 - 2\kappa t + \kappa^2 \right]^2,$$

so $\phi(t) = \dfrac{1}{(1-\kappa^2)^2} \left(g(t)\alpha + h(t)\beta \right)$.

Case $\kappa < 1$. Suppose $\kappa + \epsilon \leq t \leq 1$. We have $g'(t) = -4\kappa \left[\dfrac{2\kappa}{t} - (1+\kappa^2) \right] \dfrac{1}{t^2}$, so $g'(t) > 0$ for $t > \dfrac{2\kappa}{1+\kappa^2}$, and $g'(t) < 0$ for $t < \dfrac{2\kappa}{1+\kappa^2}$. Since $\kappa < 1$, we have $\kappa + \epsilon < \dfrac{2\kappa}{1+\kappa^2} < 1$ for $\epsilon > 0$ small. Therefore, g decreases in the interval $[\kappa + \epsilon, \dfrac{2\kappa}{1+\kappa^2}]$, and g increases in the interval $[\dfrac{2\kappa}{1+\kappa^2}, 1]$. Hence

$$\max_{[\kappa+\epsilon,1]} g(t) = \max\{g(\kappa+\epsilon), g(1)\}.$$

We have that $g(1) = (1-\kappa)^4$, and $g(\kappa+\epsilon) > g(1)$ for ϵ small, so

$$\max_{[\kappa+\epsilon,1]} g(t) = g(\kappa+\epsilon).$$

On the other hand, $h'(t) = -4\kappa \left[1 - 2\kappa\, t + \kappa^2\right]$, and so $h'(t) > 0$ for $t > \dfrac{1 + \kappa^2}{2\kappa}$
and $h'(t) < 0$ for $t < \dfrac{1 + \kappa^2}{2\kappa}$. Since $\dfrac{1 + \kappa^2}{2\kappa} > 1$, the function h is decreasing in the interval $[\kappa + \epsilon, 1]$ and so

$$\max_{[\kappa+\epsilon,1]} h(t) = h(\kappa + \epsilon).$$

Therefore we obtain that

$$\max_{[\kappa+\epsilon,1]} \phi(t) \leq \frac{1}{(1 - \kappa^2)^2}\left(\alpha\, g(\kappa + \epsilon) + \beta\, h(\kappa + \epsilon)\right).$$

It is easy to see that

$$g(\kappa + \epsilon) < (1 - \kappa^2)^2, \quad\text{and}\quad h(\kappa + \epsilon) < (1 - \kappa^2)^2$$

and so we obtain the bound

$$\max_{[\kappa+\epsilon,1]} \phi(t) \leq C_\epsilon < 1,$$

with

$$C_\epsilon = \max\left\{\frac{g(\kappa + \epsilon)}{(1 - \kappa^2)^2}, \frac{h(\kappa + \epsilon)}{(1 - \kappa^2)^2}\right\} \tag{35}$$

independent of α and β. We notice also that $C_\epsilon \to 1$ as $\epsilon \to 0^+$, and $C_\epsilon \to \left(\dfrac{1 - \kappa}{1 + \kappa}\right)^2$ as $\epsilon \to (1 - \kappa)^-$. Also notice that the function ϕ in (34) is in general not decreasing in the interval $[\kappa + \epsilon, 1]$, that is, one can choose α close to one and β close to zero with $\alpha + \beta = 1$, so that this is the case.

Case $\kappa > 1$. For ϵ small we have $\dfrac{1}{\kappa} + \epsilon < \dfrac{2\kappa}{1 + \kappa^2} < 1$, so as before, g decreases in the interval $[\dfrac{1}{\kappa} + \epsilon, \dfrac{2\kappa}{1 + \kappa^2}]$, and g increases in the interval $[\dfrac{2\kappa}{1 + \kappa^2}, 1]$. Hence

$$\max_{[(1/\kappa)+\epsilon,1]} g(t) = \max\left\{g\left(\frac{1}{\kappa} + \epsilon\right), g(1)\right\}.$$

Since now $\kappa > 1$ we have that $g(1) < g\left(\dfrac{1}{\kappa} + \epsilon\right)$, for ϵ small, and so

$$\max_{[(1/\kappa)+\epsilon,1]} g(t) = g((1/\kappa) + \epsilon).$$

Since we always have $\dfrac{1 + \kappa^2}{2\kappa} > 1$, the function h is decreasing in the interval $[(1/\kappa) + \epsilon, 1]$ and so

$$\max_{[(1/\kappa)+\epsilon,1]} h(t) = h((1/\kappa) + \epsilon).$$

Therefore we obtain that

$$\max_{[(1/\kappa)+\epsilon,1]} \phi(t) \le \frac{1}{(1 - \kappa^2)^2} \left(\alpha\, g((1/\kappa) + \epsilon) + \beta\, h((1/\kappa) + \epsilon) \right).$$

It is clear that $g((1/\kappa) + \epsilon) < (1 - \kappa^2)^2$, and $h((1/\kappa) + \epsilon) < (1 - \kappa^2)^2$ when $0 < \epsilon < 1 - (1/\kappa)$. So we obtain the bound

$$\max_{[(1/\kappa)+\epsilon,1]} \phi(t) \le C_\epsilon < 1,$$

with

$$C_\epsilon = \max\left\{ \frac{g((1/\kappa) + \epsilon)}{(1 - \kappa^2)^2}, \frac{h((1/\kappa) + \epsilon)}{(1 - \kappa^2)^2} \right\} \tag{36}$$

independent of α and β.

5.13 Application to the Far Field Refractor Problem with Loss of Energy

We have then seen that when radiation strikes a surface separating two homogeneous media I and II with different refractive indices, part of the radiation is transmitted through medium II and another part is reflected back into medium I. In fact, from Sect. 5.11 the percentage of internally reflected energy can be conveniently written for our purposes as

$$r(x) = \frac{1}{(1 - \kappa^2)^2} \left(\left[\frac{2\kappa}{x \cdot m} - (1 + \kappa^2) \right]^2 \frac{I_\|^2}{I_\|^2 + I_\perp^2} + \left[1 - 2\kappa\, x \cdot m + \kappa^2 \right]^2 \frac{I_\perp^2}{I_\|^2 + I_\perp^2} \right)$$

where $\kappa = n_2/n_1$. Therefore the percentage of energy transmitted is $t(x) = 1 - r(x)$. Here $I_\perp$ and $I_\|$ are the coefficients of the amplitude of the incident wave, which might depend on x in a continuous way. It is important to notice that from Snell's law, $x - \kappa m = \lambda\, \nu$, where ν is unit normal to the surface at the striking point and $\lambda > 0$. This implies that the function $r(x)$ is a function only depending on x and the normal ν.

We propose the following new model that takes into account the splitting of
energy. Suppose we have $f \in L^1(\Omega)$ and $g \in L^1(\Omega^*)$, both Ω, Ω^* are domains in
the sphere in $\mathbb{R}^3$, the space with physical significance for our problem. The question
is to find a surface $\mathcal{R}$ parameterized by $\{\rho(x)x : x \in \Omega\}$ that separates media I
and II such that each ray emanating from a point source, the origin, in the direction
$x \in \Omega$ with intensity $f(x)$ is refracted into a direction $m \in \Omega^*$ and received with
intensity $g(m)$. From the Fresnel formulas a surface $\mathcal{R}$ is only able to transmit in
the direction x an amount of energy equal to

$$f(x)t_{\mathcal{R}}(x)$$

where $t_{\mathcal{R}} = 1 - r_{\mathcal{R}}$, since the amount $f(x)r_{\mathcal{R}}(x)$ is reflected back. As we said,
the function $t_{\mathcal{R}}(x)$ depends of course on the surface $\mathcal{R}$ but only through x and
the unit normal vector $\nu = \nu(x)$ at the striking point. Since we will be seeking
for refracting surfaces $\mathcal{R}$, which, in particular, are convex or concave, the normal
vector $\nu(x)$ exists for almost every direction x. Also $t_{\mathcal{R}}(x) = G(x, \nu(x))$, with a
function $G(x, x')$ continuous in $\Omega \times \Omega^*$ and so $t_{\mathcal{R}}$ is defined for a.e. direction x.
We then propose the following model: the refracting surface $\mathcal{R}$ is a solution to our
problem if

$$\int_{\mathcal{T}_{\mathcal{R}}(F)} f(x) \, t_{\mathcal{R}}(x) \, dx \geq \int_F g(m) \, dm \qquad (37)$$

for each Borel subset $F \subset \Omega^*$. Here $\mathcal{T}_{\mathcal{R}}(F)$ is the collection of all directions $x \in \Omega$
that are refracted into a direction in the set F. We prove in [6] that if $\mathcal{R}$ is a refractor,
then the function $t_{\mathcal{R}}(x)$ is continuous relative to the set $\Omega \setminus S$, where S is the set
of directions where $\mathcal{R}ho$ is not differentiable, i.e., $|S| = 0$. Therefore $t_{\mathcal{R}}(x)$ is
measurable and so (37) is well defined. Since a fraction of the energy is used in
internal reflection, to be able to transmit and receive $g(m)$ a little extra energy will
be needed at the outset. A refractor $\mathcal{R}$ will be admissible to transmit the amount g if

$$\int_{\Omega} f(x) \, t_{\mathcal{R}}(x) \, dx \geq \int_{\Omega^*} g(m) \, dm. \qquad (38)$$

Since a priori we only know f, g and not $\mathcal{R}$, we do not know if this is satisfied.
In order to make sure this is the case, we proved in Sect. 5.12, that if for example
$n_2/n_1 = \kappa < 1$, then $r(x) \leq C_\epsilon < 1$ for all $x \in \Omega$ such that $x \cdot m \geq \kappa + \epsilon$,[7]
where $\epsilon > 0$ and with C_ϵ independent of $\mathcal{R}$. So if we assume that the input energy
is sufficiently larger than the output energy, then (38) holds. More precisely, if

$$\int_{\Omega} f(x) \, dx \geq \frac{1}{1 - C_\epsilon} \int_{\Omega^*} g(m) \, dm,$$

then (38) holds.

---

[7]We recall that the physical constraint for refraction is that $x \cdot m \geq \kappa$ for $\kappa < 1$, see Lemma 2.1.

Fig. 5 Refracted and
reflected vectors

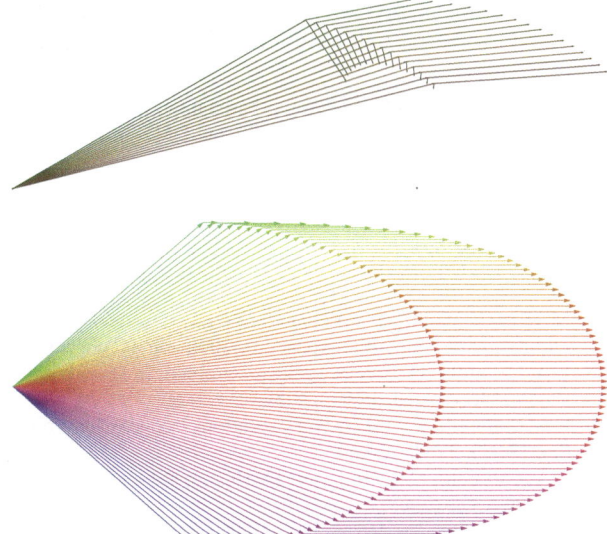

Fig. 6 Refracted vectors for
an ellipse refracting into a
fixed direction

Figure 5 represents an arc of ellipse separating glass and air, $\kappa = 2/3$, where the refracted and reflected directions are multiplied by the Fresnel coefficients $t(x)$ and $r(x)$ respectively. Figure 6 represents all the refracted vectors in an ellipse having the uniform refraction property, i.e. all rays are refracted into a fix direction, where the refracted vectors are multiplied by the Fresnel coefficient $t(x)$. Notice that the size of the refracted vectors close to the critical angle, i.e., $x \cdot m = \kappa$ tend to zero.

With this model it is proved in [6] existence of solutions, even for Radon measures μ instead of g. The basic geometry of the refractors is described in [8] and depends on κ. Indeed, the surfaces having the uniform refracting property are semi ellipsoids if $\kappa < 1$ and one sheet of hyperboloids of two sheets if $\kappa > 1$, see Sect. 2.3. A difficulty in our case is the presence of the coefficient $t_{\mathcal{R}}(x)$ in (37). This prevents us from using the optimal transportation methods used in [8]. The route used was explained in Sect. 4, that is, to solve first the problem when the right hand side is a linear combination of delta functions and then proceed by approximation. To carry out this we need to understand how the Fresnel coefficients evolve when a sequence of refractors converge. When the measure μ in the target is discrete, refractors always overshoot energy in one direction. When μ is any Radon measure, it is proved in [6] that refractors transmit more energy in any a priori chosen direction $m_o \in \overline{\Omega^*}$ which lies in the support of μ, and there is one refractor overshooting the least amount of energy in the direction of m_o. That is, for each Borel set $F \subset \Omega^*$ such that $m_o \notin F$ we have $\int_{T_{\mathcal{R}}(F)} f(x) \, t_{\mathcal{R}}(x) \, dx = \int_F g(m) \, dm$.

To place our results in perspective, we finish enumerating some related results in this area. The refractor problem assuming energy conservation, i.e., $t_{\mathcal{R}}(x) = 1$ in (37), was considered for the first time in [8] for the far field case, and in [7] and

[9] for the near field case. For reflectors also assuming energy conservation, see [2–4, 15] for the far field problem, and [11], and [10] for the near field.

Acknowledgements This material is based upon work partially supported by the National Science Foundation under Grants DMS-0901430 and DMS-1201401.

References

1. M. Born, E. Wolf, *Principles of Optics, Electromagnetic Theory, Propagation, Interference and Diffraction of Light*, 7th (expanded), 2006 edn. (Cambridge University Press, London, 1959)
2. L.A. Caffarelli, Q. Huang, Reflector problem in $\mathbf{R}^n$ endowed with non-Euclidean norm. Arch. Ration. Mech. Anal. **193**(2), 445–473 (2009)
3. L.A. Caffarelli, V. Oliker, Weak solutions of one inverse problem in geometric optics. J. Math. Sci. **154**(1), 37–46 (2008)
4. L.A. Caffarelli, C.E. Gutiérrez, Q. Huang, On the regularity of reflector antennas. Ann. Math. **167**, 299–323 (2008)
5. C.E. Gutiérrez, *The Monge–Ampère Equation* (Birkhäuser, Boston, 2001)
6. C.E. Gutiérrez, H. Mawi, The far field refractor with loss of energy. Nonlinear Anal. Theor. Meth. Appl. **82**, 12–46 (2013)
7. C.E. Gutiérrez, Q. Huang, The near field refractor, in *Geometric Methods in PDE's, Conference for the 65th birthday of E. Lanconelli*. Lecture Notes of Seminario Interdisciplinare di Matematica, vol. 7 (Universita' degli Studi della Basilicata, Potenza, 2008), pp. 175–188
8. C.E. Gutiérrez, Q. Huang, The refractor problem in reshaping light beams. Arch. Ration. Mech. Anal. **193**(2), 423–443 (2009)
9. C.E. Gutiérrez, Q. Huang, The near field refractor. Annales de l'Institut Henri Poincaré (C) Analyse Non Linéaire (to appear). https://math.temple.edu/~gutierre/papers/nearfield.final.version.pdf (2013)
10. A. Karakhanyan, X.-J. Wang, On the reflector shape design. J. Differ. Geom. **84**, 561–610 (2010)
11. S. Kochengin, V. Oliker, Determination of reflector surfaces from near-field scattering data. Inverse Probl. **13**, 363–373 (1997)
12. R.K. Luneburg, *Mathematical Theory of Optics* (University of California Press, Berkeley, 1964)
13. R. Schneider, *Convex Bodies: The Brunn–Minkowski Theory*. Encyclopedia of Mathematics and Its Applications, vol. 44 (Cambridge University Press, Cambridge, 1993)
14. A. Sommerfeld, *Electrodynamics*. Lectures on Theoretical Physics, vol. III (Academic, New York, 1952)
15. X.-J. Wang, On the design of a reflector antenna. Inverse Probl. **12**, 351–375 (1996)

On the Levi Monge-Ampère Equation

Annamaria Montanari

Abstract We are concerned with some notions of curvatures associated with pseudoconvexity and the Levi form as the classical Gauss and Mean curvatures are related to the convexity and to the Hessian matrix. In particular, given a prescribed non negative function K, the Levi Monge Ampère equation for the graph of a function $u : \mathbb{R}^{2n+1} \to \mathbb{R}$ is

$$\det \mathscr{L} = K(x, u)(1 + |Du|^2)^{\frac{n+1}{2}},$$

where $\mathscr{L}$ is the Levi form of the graph u and Du is the Euclidean gradient of u. More generally, we shall consider elementary symmetric functions of the eigenvalues of the Levi form $\mathscr{L}$ and we shall first show that these curvature equations contain information about the geometric feature of a closed hypersurface. Then, we shall show that the curvature operators lead to a new class of second order fully nonlinear equations whose characteristic form, when computed on generalized pseudoconvex functions, is nonnegative definite with kernel of dimension one. Thus, the equations are not elliptic at any point. However, they have the following redeeming feature: the missing ellipticity direction can be recovered by suitable commutation relations. We shall use this property to study existence, uniqueness and regularity of viscosity solutions of the Dirichlet problem for graphs with prescribed Levi curvature.

A. Montanari (✉)
Dipartimento di Matematica, Università di Bologna, piazza di Porta S.Donato 5,
40127 Bologna, Italy
e-mail: annamaria.montanari@unibo.it

L. Capogna et al., *Fully Nonlinear PDEs in Real and Complex Geometry and Optics*,
Lecture Notes in Mathematics 2087, DOI 10.1007/978-3-319-00942-1_4,
© Springer International Publishing Switzerland 2014

1   Introduction

In this note we present an introduction to the Levi curvature equations and we survey some existence, symmetry and regularity results for their solutions. Geometric theory of several complex variables leads to differential problems related to nonlinear second order PDE's of degenerate elliptic type. In particular, in looking for domains of holomorphy, fully nonlinear PDE's in nondivergence form appear, which are usually quoted as *Levi curvature equations* or *Levi-Monge-Ampère equations*. They are related to the *pseudoconvexity* as the classical Gauss curvature equations are related to the Euclidean convexity. These equations are of degenerate type but they have the following subelliptic property: when computed on strictly pseudoconvex functions, they become *elliptic along certain directions*, and the missing ones can be recovered by *commutations*. In the present paper we want to describe the state of the art of this subject. First of all, we will present the geometric arguments leading to the formal notion of Levi curvature, starting from the very beginning. Precisely, in Sect. 2 we introduce the Levi Form, which is the analogous to the Second Fundamental Form in differential geometry. The Levi form is the key differential geometric object which determines pseudoconvexity and many functions properties of a CR manifold. For example the holomorphic extendibility of CR functions. We also suggest the interested reader the monograph [15], that presents several differential geometric aspects in the theory of CR manifolds and tangential Cauchy-Riemann equations.

Pseudoconvexity is a central concept in complex analysis because it relates to the very core of holomorphic (i.e. complex analytic) functions, which is linked to power series, the identity theorem, and analytic continuation. This concept is a higher-dimensional phenomenon, because every open subset of the complex plane $\mathbb{C}$ is pseudoconvex. Let us recall that this is true also for Euclidean convexity: every open connected subset of $\mathbb{R}$ is convex. For the basic notion of several complex variables theory we refer to the monograph by S. Krantz [24] and to [35].

Let us now spend some words to the history of this concept. By following [36], pseudoconvexity has its roots in Friedrich Hartogs's surprising discovery in 1906 of a simple domain H in C^2 with the property that every function that is holomorphic on H has a holomorphic extension to a strictly larger open set (see Example 4). In dimension one there is no such phenomenon because every domain in the complex plane is trivially a domain of holomorphy (see Remark 3) and it is elementary to show that every Euclidean convex domain $D \subset \mathbb{C}^n$ is a domain of holomorphy. Hartogs's discovery raises the fundamental problem of characterizing those domains $D \subset \mathbb{C}^n$ for which holomorphic extension of all holomorphic functions on D does not hold. Such domains are called domains of holomorphy (see Definition 17).

Just a few years after Hartogs's discovery, E. E. Levi studied domains of holomorphy with differentiable boundaries and he found the following simple differential condition, which is very similar to the familiar differential characterization of Euclidean convexity. We assume that $D \cap U = \{z \in U : \rho(z) < 0\}$, where

ρ is a C^2 real-valued function with $d\rho \neq 0$ on a neighborhood U of a point of the boundary. In 1910 Levi proved that,

(a) If there exists a holomorphic function on $D \cap U$ which does not extend holomorphically to p (in particular, if D is a domain of holomorphy), then the Levi form at p is positive semidefinite.

(b) If the Levi form is positive definite, then the neighborhood U can be chosen so that $U \cap D$ is a domain of holomorphy.

Levi's results showed that the restriction of the complex Hessian to the complex tangent plane (nowadays called the Levi form) plays a fundamental role in the characterization of domains of holomorphy. To distinguish Levi's differential conditions from other formulations of pseudoconvexity, one refers to the condition in (a) as (Levi) pseudoconvexity. If the stronger version in (b) holds, one says that D is strictly or strongly pseudoconvex at p. By Levi's result, if D is strictly pseudoconvex at every boundary point, then D is locally a domain of holomorphy. For many years the wished global version remained a central open problem, known as the *Levi problem*. Solutions were finally obtained in the early 1950s by K. Oka, H. Bremermann, and F. Norguet. The general version of the solution of Levi's problem is then stated as: A domain in C^n is a domain of holomorphy if and only if it is pseudoconvex.

The extension to arbitrary domains requires an appropriate definition of pseudoconvexity. Many equivalent versions have been introduced over the years and the most elegant one is the formulation in Definition 16, which involves the notion of plurisubharmonic function, because general plurisubharmonic functions can be well approximated from above by C^2 or even C^∞ plurisubharmonic functions.

By following [5] in Sect. 2 we will first give the definition of the Levi form for an abstract CR structure, and then we will give more concrete representations of the Levi form for an immersed CR manifold. The Levi form for a real hypersurface will be discussed in some detail. Moreover, we will present the relationship between the Levi form and the second fundamental form of a real hypersurface.

Then, by using the Levi form, we introduce the notion of pseudoconvexity, which is a most central concept in modern complex analysis (see for instance the beautiful expository article by Range in [36]).

By following the expository article [25], in Sect. 3 we will show that the Levi problem and the construction of the envelopes of holomorphy lead to the notion of *Levi curvature* and to the Dirichlet problem for the *prescribed Levi curvature equation*. We will then introduce the prescribed Levi curvature equation for a real hypersurface of $\mathbb{C}^2$ and then we will discuss the solvability of Dirichlet problem for graphs in the class of viscosity solutions. The main result of this section is that different hypotheses on the zeros of the prescribed Levi curvature imply different regularity behavior of the solution. In particular, if the prescribed Levi curvature is smooth and strictly positive in [8] we proved the smooth solvability of the Dirichlet problem, while in the Levi flat case in [10] we proved a Frobenius type result.

In Sect. 4 we shall explicitly compute the Levi curvatures for a real hypersurface in $\mathbb{C}^{n+1}$. We then shall study the Dirichlet problem for the prescribed Gauss-Levi curvature equation in the class of Lipschitz continuous viscosity solutions, under very natural conditions on the prescribed curvature K and on the boundary data. To state our result we need a suitable notion of a generalized Levi pseudoconvex solution to the prescribed Gauss Levi curvature equation, which is a fully non linear and a non elliptic pde. This is provided by the notion of viscosity solution in the sense of Crandall et al. [12]. The modifications needed to adapt the theory of viscosity to equations of Levi Monge Ampère have been studied in the paper [13] and we will recall them for reader convenience. We will also use a comparison principle for viscosity solution, Theorem 14, and existence results of continuous and of Lipschitz continuous viscosity solutions of the prescribed Gauss Levi curvature equation, Theorem 15, which have been proved in [13]. The problem of classical regularity of Lipschitz continuous viscosity solutions is widely still open for $n > 1$. However, we shall show that the fully nonlinear PDE of the prescribed Gauss-Levi curvature equation has the following hypoellipticity property, which has been proved in [32]: every classical solution is smooth, if the prescribed Gauss-Levi curvature $K \neq 0$ at any point and $K \in C^{\infty}$.

In Sect. 5 we show the existence of nonsmooth viscosity solutions for the Levi Monge Ampère equation for $n > 1$. This generalizes a celebrated result of Pogorelov for the Monge Ampère equation (we refer to the book [17] for a deep study of the Monge Ampère equation in $\mathbb{R}^n$). In 1971 Pogorelov [34] showed that convex generalized solutions of the Monge Ampère equation

$$\det D^2 u = f(x) \tag{1}$$

in a domain $\Omega \subset \mathbb{R}^{n+1}$, $n > 1$, need not be of class C^2, even if f is positive and smooth. Urbas in [44] proved that this absence of classical regularity is not confined to equations of Monge Ampère type, but in fact occurs for the m-th elementary symmetric functions of the eigenvalues of the Hessian and the equation of prescribed m-th curvature, where in each case $m \geq 3$. Recently, in [18] we proved that a similar result holds for the Levi Monge Ampère equation.

In Sect. 6 we discuss symmetry properties for compact real hypersurfaces with some constant Levi curvatures. First of all we sketch the proof of the strong comparison principle proved in [31] and which leads to symmetry theorems for domains with constant curvatures. These results suggested the following question: are spheres the unique compact hypersurfaces with constant Levi curvatures? Klingenberg in [23] gave a first positive answer to this problem by showing that a compact and strictly pseudoconvex real hypersurface $M \subset \mathbb{C}^{n+1}$ is isometric to a sphere, provided that M has constant horizontal mean curvature, the Levi form is diagonal and the characteristic direction of the hypersurface is a geodesic. Later on in [29] we relaxed Klingerberg conditions and by using differential geometry techniques we proved that if the characteristic direction is a geodesic, then Klingerberg's Theorem holds for compact hypersurfaces with positive constant Levi mean curvature (see Theorem 24).

The study of surfaces in the Euclidean space with either constant Gauss curvature or constant mean curvature received in the past a great amount of attention. In 1899 Liebmann [26] proved that the spheres are the only compact surfaces in $\mathbb{R}^3$ with constant Gauss curvature. In 1952 Süss [41] extended the Liebmann result showing that a compact convex hypersurface in the Euclidean space must be a sphere, provided that for some j the j-th elementary symmetric polynomial in the principal curvatures is constant. In 1954 Hsiung [22] proved that the "convexity" assumption can be relaxed to the "star-shapedness" one. The proofs of the above results are based on the Minkowski formulae. A breakthrough for this sort of problems was made by Alexandrov [1] in 1956, who proved that the sphere is the only compact hypersurface embedded into the Euclidean space with constant mean curvature. Alexandrov method is completely different from the Liebmann-Süss method, and it is based on the moving plane technique, on the interior maximum principle for elliptic equations and on the boundary maximum principle of Hopf type for uniformly elliptic equations. In 1978 Reilly [37] obtained another proof of the Alexandrov theorem combining the Minkowski formulae with some new elegant arguments.

In Sect. 6, by following [28], we will show how to use the null Lagrangian property for elementary symmetric functions of the eigenvalues of the complex Hessian matrix and the classical divergence theorem to prove an integral formula (59) for a closed hypersurface in term of the j-th Levi curvature. Then, we will follow the Reilly approach to prove an isoperimetric inequality Theorem 22 and we will use the Minkowski formula for the classical mean curvature to conclude an Alexandrov type result.

The problem of characterizing hypersurfaces with constant Levi curvature has been recently studied by many authors. Let us mention that Hounie and Lanconelli in [20] showed the result for Reinhardt domain of $\mathbb{C}^2$, i.e. for domains D such that if $(z_1, z_2) \in D$ then $(e^{i\theta_1}z_1, e^{i\theta_2}z_2) \in D$ for all real θ_1, θ_2. Their technique has then been used in [21] to prove an Alexandrov Theorem for Reinhardt domains in $\mathbb{C}^{n+1}$ with spherical symmetry for every n. Then in [33] Monti and Morbidelli proved a Darboux-type theorem for $n \geq 2$: the unique Levi umbelical hypersurfaces in $\mathbb{C}^{n+1}$ with all constant Levi curvatures are spheres or cylinders. However, to the best of our knowledge, nowadays it is still not clear if balls are the only compact hypersurfaces with constant Gauss Levi curvature.

2   The Levi Form

In this first section we introduce the Levi Form, which is the analogous to the Second Fundamental Form in differential geometry. The Levi form is the key differential geometric object which determines many functions theoretic properties of a CR manifold. For example the holomorphic extendibility of CR functions.

By following [5] we will first give the definition of the Levi form for an abstract CR structure, and then we will give more concrete representations of the Levi form for an immersed CR manifold. The Levi form for a real hypersurface will be discussed in some detail. Moreover, we will present the relationship between the Levi form and the second fundamental form of a real hypersurface.

   We start with notation

2.1 Complexified Vectors

By starting from the very beginning, we define objects such as

$$\frac{\partial}{\partial z_j} = \frac{1}{2} \left(\frac{\partial}{\partial x_j} - i \frac{\partial}{\partial y_j} \right), \tag{2}$$

which are often encountered in complex analysis. This is a complexified vector due to presence of $i = \sqrt{-1}$.

   We now make this notion precise.

   Suppose V is a real vector space. The complexification of V is the tensor product

$$V \otimes \mathbb{C} = \{X + iY; \ X, Y \in V\} \tag{3}$$

If V is a vector space of real dimension N then $dim_{\mathbb{R}} V \otimes \mathbb{C} = 2N, dim_{\mathbb{C}} V \otimes \mathbb{C} = N$.

2.2 Complex Structure

Definition 1. Suppose V is a real vector space. A linear map $J : V \rightarrow V$ is a complex structure if $J \circ J = -Id$.

A complex structure can only be defined on an even-dimensional real vector space because $(\det J)^2 = (-1)^N$, with $N = dim_{\mathbb{R}} V$.

2.2.1 Standard Complex Structure

Example 1. Let $V = T_p(\mathbb{R}^{2n})$ be the space of vector fields in $\mathbb{R}^{2n}$. We give $\mathbb{R}^{2n}$ the coordinates $(x_1, y_1, \ldots, x_n, y_n)$. The standard complex structure J for $T_p(\mathbb{R}^{2n})$ is defined by setting for all $j = 1, \ldots, n$

$$J \left(\frac{\partial}{\partial x_j} \right) = \frac{\partial}{\partial y_j} \quad J \left(\frac{\partial}{\partial y_j} \right) = -\frac{\partial}{\partial x_j} \tag{4}$$

and then by extending J to all $T_p(\mathbb{R}^{2n})$ by linearity.

This complex structure map is designed to simulate multiplication by $i = \sqrt{-1}$.
   It is an isometry with respect to the Euclidean metric on $\mathbb{R}^{2n}$.

2.2.2 Eigenvalues and Eigenspaces of the Standard Complex Structure

Since $J \circ J = -Id$ on V, the same holds true on $V \otimes \mathbb{C}$. Therefore, $J : V \otimes \mathbb{C} \to V \otimes \mathbb{C}$ has eigenvalues $+i$ and $-i$ with corresponding eigenspaces denoted by $V^{1,0}$ and $V^{0,1}$. From elementary linear algebra we have $V \otimes \mathbb{C} = V^{1,0} \oplus V^{0,1}$.

As an example, let $V = T_p(\mathbb{R}^{2n}) \simeq T_p(\mathbb{C}^n)$. Define the vector fields

$$\frac{\partial}{\partial z_j} = \frac{1}{2} \left(\frac{\partial}{\partial x_j} - i \frac{\partial}{\partial y_j} \right) \qquad \frac{\partial}{\partial \bar{z}_j} = \frac{1}{2} \left(\frac{\partial}{\partial x_j} + i \frac{\partial}{\partial y_j} \right) \tag{5}$$

A basis for $T_p^{1,0}(\mathbb{C}^n)$ is given by $\{\partial_{z_1}, \ldots, \partial_{z_n}\}$ and a basis for $T_p^{0,1}(\mathbb{C}^n)$ is given by $\{\partial_{\bar{z}_1}, \ldots, \partial_{\bar{z}_n}\}$.

2.3 Immersed CR Manifold

We now give the definition of an immersed CR manifold, which is the simplest case of CR manifold. For a smooth submanifold M of $\mathbb{C}^n$, we denote by $T_p(M)$ the real tangent space of M at a point $p \in M$.

2.3.1 Holomorphic Tangent Space

In general $T_p(M)$ is not invariant under the standard complex structure J for $T_p(\mathbb{C}^n) \simeq T_p(\mathbb{R}^{2n})$. Therefore we give special designation to the largest J-invariant subspace of $T_p(M)$.

Definition 2. For a point $p \in M$, the complex tangent space of M at p is the vector space

$$H_p(M) = T_p(M) \cap J(T_p(M)). \tag{6}$$

Remark that $H_p(M)$ must be an even-dimensional real vector space, because $J \circ J|_{H_p(M)} = -Id$ and therefore $|\det J|_{H_p(M)}|^2 = (-1)^m$ with $m = \dim H_p(M)$.

2.3.2 Totally Real Part of the Tangent Space

We also give special designation for the other directions in $T_p(M)$ which do not lie in $H_p(M)$.

Definition 3. The totally real part of the tangent space of M is the quotient space

$$X_p(M) = T_p(M)/H_p(M). \tag{7}$$

Using the Euclidean inner product on $T_p(\mathbb{R}^{2n})$ we can identify $X_p(M)$ with the orthogonal complement of $H_p(M)$ and we have $T_p(M) = H_p(M) \oplus X_p(M)$.

Remark that $J(X_p(M))$ is orthogonal to $H_p(M)$. Therefore, it is transverse to $T_p(M)$.

From elementary linear algebra we have

Lemma 1. *Suppose M is a real submanifold of $\mathbb{C}^n$ of real dimension $2n - d$, then*

$$2n - 2d \leq dim_{\mathbb{R}} H_p(M) \leq 2n - d, \quad 0 \leq dim_{\mathbb{R}} X_p(M) \leq d. \tag{8}$$

The dimensions of $H_p(M)$ and of $X_p(M)$ are of crucial importance, as we will see in a moment.

If M is a real hypersurface, then $d = 1$, and the only possibility is $dim_{\mathbb{R}} H_p(M) = 2n - 2$. In particular the dimension of $H_p(M)$ never changes. If $d > 1$ there are more possibilities.

2.3.3 Immersed CR Manifold

Definition 4. A submanifold M of $\mathbb{C}^n$ is called an immersed CR manifold if $dim_{\mathbb{R}} H_p(M)$ is independent of $p \in M$.

Since $H_p(M)$ is J-invariant, the complex structure map J on $T_p(\mathbb{R}^{2n}) \otimes \mathbb{C} = \{X + iY : X, Y \in T_p(\mathbb{R}^{2n})\}$ restricts to a complex structure on $H_p(M) \otimes \mathbb{C}$ and $H_p(M) \otimes \mathbb{C}$ admits a decomposition as direct sum of $H_p^{1,0}(M)$ and $H_p^{0,1}(M)$, which are the $+i$ and $-i$ eigenspaces of J, respectively.

It will be useful to have a way of identifying the above spaces in terms of a local defining system for M.

Lemma 2. *Let M be a smooth manifold of $\mathbb{C}^n$ defined near a point $p \in M$ by $M = \{z \in \mathbb{C}^n : \rho_1(z) = \cdots \rho_d(z) = 0\}$ where $\rho_1, \cdots \rho_d$ are smooth real valued functions with $d\rho_1(z) \wedge \cdots \wedge d\rho_d(z) \neq 0$ near p.*

$$W = \sum_{j=1}^{n} w_j \partial_{z_j} \in H_p^{1,0}(M) \iff W(\rho_k)(p) = 0, 1 \leq k \leq d$$

$$\overline{W} = \sum_{j=1}^{n} w_j \partial_{\bar{z}_j} \in H_p^{0,1}(M) \iff \overline{W}(\rho_k)(p) = 0, 1 \leq k \leq d. \tag{9}$$

Definition 5. We say that a subbundle V is involutive if it is closed under the Lie bracket, that is $[X_1, X_2]$ belongs to V whenever $X_1, X_2 \in V$.

Lemma 3. *Suppose M is a CR submanifold of $\mathbb{C}^n$. Then*

- $H_p^{1,0}(M) \cap H_p^{0,1}(M) = \{0\}$ *for each $p \in M$*
- $H_p^{1,0}(M)$ *and* $H_p^{0,1}(M)$ *are involutive.*

2.4   CR Structure

Let $T^{\mathbb{C}}(M)$ denote the complexified tangent bundle whose fibers at each point $p \in M$ are $T_p(M) \otimes \mathbb{C}$.

Definition 6. Let M be a smooth manifold and suppose $\mathbb{L}$ is a subbundle of $T^{\mathbb{C}}(M)$. The pair $(M, \mathbb{L})$ is called a CR structure (or an abstract CR manifold) if

- $\mathbb{L}_p \cap \overline{\mathbb{L}}_p = \{0\}$ for each $p \in M$
- $\mathbb{L}$ is involutive.

It is clear from the above discussion that if M is a CR submanifold of $\mathbb{C}^n$ then the pair $(M, H^{1,0}(M))$ is a CR structure.

2.4.1   The Levi Form for a CR Structure

Given a CR structure $(M, \mathbb{L})$ the subbundle $\mathbb{L} \oplus \overline{\mathbb{L}} \subset T^{\mathbb{C}}(M)$ is not necessarily involutive. The Levi form for M is defined so that it measures the degree to which $\mathbb{L} \oplus \overline{\mathbb{L}}$ fails to be involutive.

For $p \in M$ let $\pi_p : T_p(M) \otimes \mathbb{C} \to (T_p(M) \otimes \mathbb{C})/(\mathbb{L} \oplus \overline{\mathbb{L}})$ be the natural projection map.

Definition 7. The Levi form at a point $p \in M$ is the map $\mathscr{L}_p : \mathbb{L} \times \overline{\mathbb{L}} \to (T_p(M) \otimes \mathbb{C})/(\mathbb{L} \oplus \overline{\mathbb{L}})$ defined by

$$\mathscr{L}_p(L_p, \overline{W}_p) = \frac{1}{2i}\pi_p([L_p, \overline{W}_p]), \quad \forall\, L_p \in \mathbb{L}_p, \overline{W}_p \in \overline{\mathbb{L}}_p \tag{10}$$

where $L, \overline{W}$ are any vector fields in $\mathbb{L}, \overline{\mathbb{L}}$ respectively, which equal $L_p, \overline{W}_p$ at p.

Remark 1. Let us remark the following facts:

- The definition is independent of the $\mathbb{L}$ vector field extension of the vector $L_p \in \mathbb{L}_p$
- The vector field $[L, \overline{W}]$ lies in $T^{\mathbb{C}}(M)$ because $T^{\mathbb{C}}(M)$ is involutive.
- The Levi Form measures the piece of $\frac{1}{2i}([L, \overline{W}]_p)$ that lies outside of $\mathbb{L}_p \oplus \overline{\mathbb{L}}_p$.
- The factor $\frac{1}{2i}$ is introduced to make the Levi form real valued as a quadratic form, i.e. $\overline{\mathscr{L}_p(L_p, \overline{L}_p)} = \mathscr{L}_p(L_p, \overline{L}_p)$, for all $L_p \in \mathbb{L}_p$.

2.4.2   Levi Flat

Definition 8. We say that a CR structure $(M, \mathbb{L})$ is Levi Flat if the Levi form of M vanishes at each point in M.

Example 2. Let $M = \{(z, w) \in \mathbb{C} \times \mathbb{C}^{n-1} : Im\, z = 0\}$. A basis for $\mathbb{L} = H^{1,0}(M)$ is given by $\frac{\partial}{\partial w_1}, \ldots, \frac{\partial}{\partial w_{n-1}}$. Since $\left[\frac{\partial}{\partial w_j}, \frac{\partial}{\partial \overline{w}_k}\right] = 0$ then M is Levi Flat.

Also note that M is foliated by the complex manifolds

$$M_x = \{(x, w) \in \mathbb{C} \times \mathbb{C}^{n-1}\}, \quad x \in \mathbb{R}. \tag{11}$$

The complexified tangent bundle at each M_x is given by $\mathbb{L} \oplus \overline{\mathbb{L}}$.

Theorem 1. *Suppose $(M, \mathbb{L})$ is a Levi Flat CR structure. Then M is locally foliated by complex manifold whose complexified tangent bundle is given by $\mathbb{L} \oplus \overline{\mathbb{L}}$.*

The idea of the proof is to prove that $\mathbb{L} \oplus \overline{\mathbb{L}}$ and its underlying real bundle $H(M) = \{L + \overline{L} : L \in \mathbb{L}\}$ are involutive. Then the foliation is obtained by the real Frobenius Theorem.

2.5 The Levi Form for an Immersed CR Manifold

For an immersed CR manifold M

$$\mathbb{L} = H^{1,0}(M), \overline{\mathbb{L}} = H^{0,1}(M), \tag{12}$$

and the quotient space $T^{\mathbb{C}}(M)/\mathbb{L} \oplus \overline{\mathbb{L}}$ is identified with the totally real part of the real tangent space.

Definition 9. The Levi form of M at $p \in M$ is the map

$$\mathscr{L}_p : H^{1,0}(M) \times H^{0,1}(M) \to X_p(M) \tag{13}$$

$$\mathscr{L}_p(L_p, \overline{W}_p) = \frac{1}{2i}\pi_p[L_p, \overline{W}_p], \quad L_p \in H^{1,0}(M), \overline{W}_p \in H^{0,1}(M) \tag{14}$$

where $\pi_p : T_p(M) \to X_p(M)$ is the orthogonal projection.

2.6 Levi Form of a Real Hypersurface

Let M be a real hypersurface in $\mathbb{C}^n$ that separates $\mathbb{R}^{2n}$ in two open sets D and $\mathbb{R}^{2n} \setminus D$. Since $dim\, X_p(M) = 1$ we can forget about the vectorial structure of the Levi form.

Let $\mathscr{N}_p$ denote the unit inward normal to D at $p \in M$.

Definition 10. $T_p = J(\mathscr{N}_p)$ is called the characteristic direction of M at p.

We have $Span_{\mathbb{R}}\{T_p\} = X_p(M)$.

Definition 11. The Levi form of M at $p \in M$ is the $\mathbb{C}$-bilinear Hermitian form $\mathscr{L}_p : H^{1,0}(M) \times H^{0,1}(M) \rightarrow \mathbb{C}$, $\mathscr{L}_p(L_p, \overline{W}_p) = \frac{1}{2i}\langle[L_p, \overline{W}_p], T_p\rangle$, $L_p \in H^{1,0}(M)$, $\overline{W}_p \in H^{0,1}(M)$ where $\langle\cdot,\cdot\rangle$ is the $\mathbb{C}$-linear extension of the Euclidean inner product.

Recall that $\partial_{z_j} = \frac{1}{2}(\partial_{x_j} - i\,\partial_{y_j})$, $\partial_{\bar{z}_k} = \frac{1}{2}(\partial_{x_k} + i\,\partial_{y_k})$ and

$$\langle\partial_{z_j}, \partial_{\bar{z}_k}\rangle = \frac{1}{2}\delta_{jk}, \quad \langle\partial_{z_j}, \partial_{z_k}\rangle = 0. \tag{15}$$

2.7 Comparison with the Second Fundamental Form

To compare the Levi Form with the Second fundamental form it is convenient to introduce the Levi Civita connection.

Definition 12 (Levi Civita connection). Suppose $W = \sum w_j \partial_{x_j}$ is a vector field on $\mathbb{R}^N$ and let $V_p \in T_p(\mathbb{R}^N)$. Define $\nabla_{V_p} W \in T_p(\mathbb{R}^N)$

$$\nabla_{V_p} W = \sum_{j=1}^{N} V_p(w_j)\partial_{x_j}. \tag{16}$$

Remark that $\nabla_{V_p} W$ is the derivative of W in the direction of V_p and that $[V, W]_p = \nabla_{V_p} W - \nabla_{W_p} V$.

Definition 13. The second fundamental form is the map $\mathbb{III}_p(M) : T_p(M) \times T_p(M) \rightarrow \mathbb{R}$ defined by

$$\mathbb{III}_p(V_p.W_p) = \langle -\nabla_{V_p}\mathcal{N}, W_p\rangle, \quad V_p.W_p \in T_p(M)$$

where $\langle\cdot,\cdot\rangle$ is the Euclidean inner product.

We now derive a formula for $\mathbb{III}_p$ in term of the real Hessian of the following defining function for M

$$\rho(x) = \begin{cases} -dist(x, M) & \text{if } x \in D \\ dist(x, M) & \text{if } x \in \mathbb{R}^N \setminus D \end{cases} \tag{17}$$

If M is $C^k, k \geq 2$ then ρ is C^k near M in $\mathbb{R}^N$ and $-\nabla\rho = \mathcal{N}$ on M.
Let $V_p = \sum v_j \partial_{x_j}, W_p = \sum w_k \partial_{x_k} \in T_p(M)$. Then

$$\mathbb{III}(V_p, W_p) = \sum \frac{\partial^2 \rho(p)}{\partial x_j \partial x_k} v_j w_k \tag{18}$$

Moreover, we have

$$\mathbb{III}_p(V_p, W_p) = \langle \nabla_{V_p} W, \mathcal{N}_p \rangle \tag{19}$$

We now extend the Levi Civita connection to $\mathbb{C}^n$ by $\mathbb{C}$-linearity and we get

Theorem 2. *For all* $Z \in H^{1,0}(M), \overline{W} \in H^{0,1}(M)$

$$\mathscr{L}(Z, \overline{W}) = \frac{1}{2i} \langle [Z, \overline{W}], T \rangle = \langle \nabla_Z \overline{W}, \mathcal{N} \rangle = \langle -\nabla_Z \mathcal{N}, \overline{W} \rangle \tag{20}$$

Proof. By using the fact that $[Z, \overline{W}] = \nabla_Z \overline{W} - \nabla_{\overline{W}} Z$ is a tangent vector fields, we get

$$\langle [Z, \overline{W}], T \rangle = \langle \nabla_Z \overline{W} - \nabla_{\overline{W}} Z, T \rangle = \langle \nabla_Z J(\overline{W}) - \nabla_{\overline{W}} J(Z), J(T) \rangle \tag{21}$$

$$= i \langle \nabla_Z \overline{W} + \nabla_{\overline{W}} Z, \mathcal{N} \rangle = 2i \langle \nabla_Z \overline{W}, \mathcal{N} \rangle = 2i \langle -\nabla_Z \mathcal{N}, \overline{W} \rangle$$

$\square$

Theorem 3. *For all* $Z = X - iY \in H^{1,0}(M)$ *with* $Y = J(X)$, *we have*

$$\mathscr{L}(Z, \overline{Z}) = \mathbb{III}(X, X) + \mathbb{III}(Y, Y) \tag{22}$$

Proof. Since $[X, Y] = \nabla_X Y - \nabla_Y X$ we have

$$\mathscr{L}(Z, \overline{Z}) = \frac{1}{2i} \langle [Z, \overline{Z}], T \rangle = \frac{1}{2i} \langle [X - iY, X + iY], T \rangle = \langle [X, Y], T \rangle \tag{23}$$

$$= \langle \nabla_X Y - \nabla_Y X, T \rangle = \langle \nabla_X J(Y) - \nabla_Y J(X), J(T) \rangle$$

$$= \langle -\nabla_X X - \nabla_Y Y, -\mathcal{N} \rangle = \langle \nabla_X X, \mathcal{N} \rangle + \langle \nabla_Y Y, \mathcal{N} \rangle.$$

$\square$

Theorem 4 (LEVI FORM OF A REAL HYPERSURFACE). *Let* $M = \{z \in \mathbb{C}^n : \rho(z) = 0\}$ *be a smooth real oriented hypersurface. If* $p \in M$ *and* $\partial \rho(p) = (\partial_{z_1} \rho(p), \ldots, \partial_{z_n} \rho(p)) \neq 0$, *then*

$$\mathscr{L}_p(V, \overline{W}) = \frac{1}{2|\partial \rho(p)|} \sum_{j,k=1}^{n} \frac{\partial^2 \rho}{\partial z_j \partial \overline{z}_k}(p) v_j \bar{w}_k \tag{24}$$

for $V = \sum_{k=1}^{n} v_k \partial_{z_k}, W = \sum_{k=1}^{n} w_k \partial_{z_k} \in H_p^{1,0}(M).$

Proof. The inward unit normal vector is $\mathcal{N} = -\frac{\rho_{\overline{z}_\ell} \partial_{z_\ell} + \rho_{z_\ell} \partial_{\overline{z}_\ell}}{|\partial \rho|} = -\frac{\rho_{x_\ell} \partial_{x_\ell} + \rho_{y_\ell} \partial_{y_\ell}}{|\nabla \rho|}$

$$\mathscr{L}_p(V,\overline{W}) = \langle -\nabla_V \mathscr{N}, \overline{W} \rangle = \sum_{j,k=1}^{n} v_j \bar{w}_k \langle \partial_{z_j} \left(\frac{\rho_{\bar{\ell}}}{|\partial\rho|} \right) \partial_{z_\ell}, \partial_{\bar{z}_k} \rangle \qquad (25)$$

$$= \frac{1}{2} \sum_{j,k=1}^{n} v_j \bar{w}_k \partial_{z_j} \left(\frac{\rho_{\bar{k}}}{|\partial\rho|} \right) = \frac{1}{2} \sum_{j,k=1}^{n} v_j \bar{w}_k \frac{\partial_{z_j} \rho_{\bar{k}}}{|\partial\rho|} \qquad \square$$

For this reason the Levi form is identified with the restriction of the complex Hessian of ρ to $H_p^{1,0}(M)$.

It is important to note that if $\tilde{\rho}$ is another defining function for M with $d\tilde{\rho} \neq 0$ on M then the map

$$(v_1, \ldots, v_n, \bar{w}_1, \ldots, \bar{w}_n,) \in H_p^{1,0}(M) \times H_p^{0,1}(M) \mapsto \sum_{j,k=1}^{n} \frac{\partial^2 \tilde{\rho}}{\partial z_j \, \partial \bar{z}_k}(p) v_j \bar{w}_k \quad (26)$$

is a nonzero multiple of the Levi Form at p. To see this, write $\tilde{\rho} = \alpha\rho$ for some smooth function $\alpha : \mathbb{C}^n \to \mathbb{R}$ which is nonzero near M.

2.7.1   Levi Form of a Real Hypersurface in $\mathbb{C}^2$

Example 3. Let $Z = \rho_{z_2}\partial_{z_1} - \rho_{z_1}\partial_{z_2}$. Since $Z(\rho) = 0$ we have $Z \in H^{1,0}(M)$ and

$$\mathscr{L}(Z,\overline{Z}) = \frac{1}{2|\partial\rho|} \left(|\rho_{z_2}|^2 \rho_{z_1\bar{z}_1} - \rho_{z_1} \rho_{\bar{z}_2} \rho_{z_2\bar{z}_1} - \rho_{z_2} \rho_{\bar{z}_1} \rho_{z_1\bar{z}_2} + |\rho_{z_1}|^2 \rho_{z_2\bar{z}_2} \right)$$

$$= \frac{-1}{|\nabla\rho|} \det \begin{pmatrix} 0 & \rho_{z_1} & \rho_{z_2} \\ \rho_{\bar{z}_1} & \rho_{z_1\bar{z}_1} & \rho_{z_2\bar{z}_1} \\ \rho_{\bar{z}_2} & \rho_{z_1\bar{z}_2} & \rho_{z_2\bar{z}_2} \end{pmatrix}. \qquad (27)$$

Here ∇ is the Euclidean gradient vector.

2.8   Strictly Pseudoconvexity

Definition 14. An oriented real hypersurface M is called strictly pseudoconvex at a point $p \in M$ if the Levi form at p is positive definite, i.e. if there exists a defining function ρ for M so that

$$\sum_{j,k=1}^{n} \frac{\partial^2 \rho}{\partial z_j \, \partial \bar{z}_k}(p) w_j \bar{w}_k > 0 \qquad (28)$$

for all $W = \sum_{k=1}^{n} w_k \partial_{z_k} \in H_p^{1,0}(M)$.

The above inequality is invariant under biholomorphic change of coordinates. This follows by explicitly computing the complex Hessian of $\rho \circ F$ where F is a biholomorphism.

Strictly pseudoconvex hypersurfaces have the following beautiful property

Theorem 5 (Narasimhan). *Suppose M is strictly pseudoconvex at a point $p \in M$. Then there is a biholomorphic map F defined in a neighborhood U of p so that $F(M \cap U)$ is strictly convex in $F(U)$.*

It is elementary, but nontrivial, to show that a domain is strictly pseudoconvex near p if and only if it is strictly Euclidean convex (i.e., the relevant matrix of second-order partial derivatives is positive definite) with respect to suitable local holomorphic coordinates centered at p. Stated differently, strict pseudoconvexity is locally simply the biholomorphically invariant version of strict convexity. Unfortunately, this neat characterization breaks down if the Levi form is only positive semidefinite for each $p \in M$, as shown by an example discovered by J. J. Kohn and L. Nirenberg in 1972.

2.9 Pseudoconvex Domains

Definition 15. The domain $D = \{z \in \mathbb{C}^n : f(z) < 0\}$, with $f \in C^2$, is called *pseudoconvex* if

$$\mathscr{L}_p \geq 0, \quad \forall \, p \in \partial D.$$

An equivalent definition (see for instance [24] for the proof), which will be useful in the sequel, is the following :

Definition 16. D is pseudoconvex if

$$D \ni z \mapsto -\log(dist(z, \partial D)) \quad \text{is } plurisubharmonic.$$

Remark 2. This last definition does not require any regularity of ∂D and it is called Hartog's pseudoconvexity. Note that it is a continuous function which tends to $+\infty$ as $z \to \partial D$.

One verifies that convex domains are pseudoconvex and that a domain with C^2 boundary is pseudoconvex according to this definition if and only if it is Levi pseudoconvex. Also, any pseudoconvex domain is the increasing union of strictly (Levi) pseudoconvex domains with C^∞ boundaries. Let us recall that a domain D is (Euclidean) convex if and only if $-\log dist(z, \partial D)$ is a convex function.

3   The Levi Problem and Levi Curvature for a Real Hypersurface

In this section we will talk about the Levi problem, which leads to the notion of *Levi curvature* and to the Dirichlet problem for the *prescribed Levi curvature equation*.

We will introduce the prescribed Levi curvature equation for a real hypersurface of $\mathbb{C}^2$ and then we will discuss the solvability of Dirichlet problem for graphs in the class of viscosity solutions. The main result of this section is that different hypotheses on the zeros of the prescribed Levi curvature imply different regularity behavior of the solution.

Let us begin with some classical things about domains of holomorphy.

Definition 17. $D \subset \mathbb{C}^n$ is a domain of holomorphy if for every $p \in \partial D$ there exists

$$F_p : D \to \mathbb{C}, \quad \text{holomorphic}$$

which does not extend to any neighborhood of p

Remark 3. Every domain $D \subset \mathbb{C}$ is a domain of holomorphy: just take

$$F_p(z) = \frac{1}{z - p}.$$

It is clearly holomorphic in D, but surely it has no holomorphic extension to any neighborhood of p. This sort of simple construction does not extend to more than one variable, as the zeroes and singularities of holomorphic functions are not isolated in the case of two or more variables.

Example 4 (Hartogs's hammer (1906)). There exists a domain $D \subset \mathbb{C}^2$ which is not a domain of holomorphy:

$$H = \left\{ (z_1, z_2) \in C^2 : |z_1| < \frac{1}{2}, |z_2| < 1 \right\} \cup \left\{ |z_1| < 1, \frac{1}{2} < |z_2| < 1 \right\}$$

Every holomorphic function $F : H \to \mathbb{C}$ can be extended to a holomorphic function $F : \tilde{H} \to \mathbb{C}$, where

$$\tilde{H} = \{|z_1| < 1, |z_2| < 1\} \quad \text{strictly contains} \quad H$$

Proof. Let $f : H \to \mathbb{C}$ be holomorphic and fix r with $1/2 < r < 1$. The function

$$F(z_1, z_2) = \frac{1}{2\pi i} \int_{|w|=r} \frac{f(z_1, w)}{w - z_2} dw$$

is easily seen to be holomorphic on $G = \{|z_1| < 1, |z_2| < r\}$. Observe that for fixed $\tilde{z}_1$ with $|\tilde{z}_1| < 1/2$ the function $z_2 \to f(\tilde{z}_1, z_2)$ is holomorphic on the disc $\{|z_2| < 1\}$

and hence by the Cauchy integral formula $f(\tilde{z}_1, z_2) = F(\tilde{z}_1, z_2)$ for $|z_2| < r$. Thus $f \equiv F$ on $\{|z_1| < 1/2, |z_2| < r\}$, which implies $f \equiv F$ on $H \cap G$ by the identity theorem, so that F provides the holomorphic extension of f from H to $\tilde{H} = H \cup G$.   □

In 1910 E.E. Levi tried to characterize domains of holomorphy in term of a differential property of the boundary. He showed that every domain of holomorphy is pseudoconvex and that, if $p \in \partial D$ and $\mathcal{L}_p > 0$, then $D \cap U$ is a *domain of holomorphy*, for a suitable neighborhood U of p.

Comparison between pseudoconvexity and domains of holomorphy was called the *Levi problem* and it was completely solved in 1954 by Oka, Bremmerman and Norgouet:

$$D \subset \mathbb{C}^n \text{ is a domain of holomorphy iff } D \text{ is pseudoconvex.}$$

A natural subsequent problem is the following: given $D \subset \mathbb{C}^n$ look for its *holomorphic hull*, i.e, the smallest domain of holomorphy containing D.

This problem led to the notion of *Levi curvature* and to the Dirichlet problem for the *prescribed Levi curvature equation*.

3.1 Levi Curvature for a Real Hypersurface in $\mathbb{C}^2$

Let $D = \{(z_1, z_2) \in \mathbb{C}^2 : f(z_1, z_2) < 0\}$, $\partial_p f \neq 0$ if $f(p) = 0$. Consider the Levi form of $M = \partial D$ at a point $p \in \partial D$

$$\mathcal{L}(Z, \overline{Z}) \text{ for } Z \in H_p^{1,0}(\partial D)$$

Definition 18. We define the *Levi curvature* of ∂D at the point p as

$$K_{\partial D}(p) := \mathcal{L}(Z, \overline{Z}), \qquad Z \in H_p^{1,0}(\partial D), \ \langle Z, \overline{Z} \rangle = 1$$

We can explicitly write it in term of a defining function f for the domain D. Indeed, take $Z = \frac{\sqrt{2}}{|\partial f|}(f_2 \partial_1 - f_1 \partial_2)$. Then $\langle Z, \overline{Z} \rangle = 1$ and

$$K_{\partial D}(p) = \mathcal{L}(Z, \overline{Z}) = \frac{\Delta_p(f)}{|\partial_p f|^3}$$

where

$$\Delta_p(f) = (f_{z_2} f_{\overline{z_2}}) f_{z_1 \overline{z_1}} - (f_{z_2} f_{\overline{z_1}}) f_{z_1 \overline{z_2}} - (f_{z_1} f_{\overline{z_2}}) f_{z_2 \overline{z_1}} + (f_{z_1} f_{\overline{z_1}}) f_{z_2 \overline{z_2}}$$

The *Levi curvature* of ∂D was first introduced by Bedford-Gaveau [2] and Tomassini [42] by analogy with the real case. However, a posteriori we are able to recognize its geometric meaning of complex curvature.

Remark 4. Let us remark that

- $K_{\partial D}(p)$ is real and independent of the defining function f of D
- $K_{\partial D}(p)$ is invariant with respect to unitary biholomorphic transformation of $\mathbb{C}^2$

3.1.1 Levi Curvature of a Ball

Example 5. Let $B_R = \{z = (z_1, z_2) \in \mathbb{C}^2 : |z| < R\}$. Then at every $p \in \partial B_R$

$$K_{\partial B_R}(p) = \frac{1}{R}.$$

Proof. We can choose $f = |z_1|^2 + |z_2|^2 - R^2$ as a defining function of B_R. We have

$$|\partial f|^2 = |z_1|^2 + |z_2|^2 = R^2$$
$$\Delta_p(f) = (f_{z_2} f_{\overline{z_2}}) f_{z_1 \overline{z_1}} - (f_{z_2} f_{\overline{z_1}}) f_{z_1 \overline{z_2}} - (f_{z_1} f_{\overline{z_2}}) f_{z_2 \overline{z_1}} + (f_{z_2} f_{\overline{z_1}}) f_{z_2 \overline{z_2}}$$
$$= |z_2|^2 + |z_1|^2 = R^2$$
$$K_{\partial B_R}(p) = \frac{\Delta_p(f)}{|\partial_p f|^3} = \frac{R^2}{R^3} = \frac{1}{R} \qquad \qquad \square$$

3.1.2 Levi Curvature of a Graph in $\mathbb{C}^2$

Let $u : \Omega \to \mathbb{R}$ be a C^2-function on the open set $\Omega \subset \mathbb{R}^3$.

The graph of u, $M := \{(x, y, t, \tau) \in \mathbb{R}^4 : \tau = u(x, y, t)\}$ is part of the boundary of

$$D = \{(x, y, t, \tau) \in \mathbb{R}^4 : \tau > u(x, y, t)\}$$

We identify $\mathbb{R}^4$ with $\mathbb{C}^2$ and denote $z \in \mathbb{C}^2$ as

$$z = (z_1, z_2), \ z_1 = x + iy, \ z_2 = t + i\tau.$$

Then $D = \{(z_1, z_2) \in \mathbb{C}^2 : f(z_1, z_2) < 0\}$, with

$$f(z_1, z_2) = f(x, y, t, \tau) := u(x, y, t) - \tau$$

so that

$$f_{z_1} = \frac{1}{2}(f_x - if_y) = \frac{1}{2}(u_x - iu_y) \text{ and } f_{z_2} = (f_t - if_\tau) = \frac{1}{2}(u_t + i).$$

At a point $p = (\eta, u(\eta)) \in M$, $\eta = (x, y, t) \in \Omega$, one has

$$K_M(p) = \frac{\Delta_p(f)}{|\partial_p f|^3} = \mathscr{L}_0(u) \frac{1 + u_t^2}{(1 + |Du|^2)^{\frac{3}{2}}}, \quad D = (\partial_x, \partial_y, \partial_t)$$

where

$$\mathscr{L}_0(u) := u_{xx} + u_{yy} + 2a\, u_{xt} + 2b\, u_{yt} + (a^2 + b^2) u_{tt}$$

and

$$a = a(Du) = \frac{u_y - u_x u_t}{1 + u_t^2}, \quad b = b(Du) = -\frac{u_x + u_y u_t}{1 + u_t^2}.$$

Then, given a function $K = K(\eta, r)$, the PDE

$$\mathscr{L}_0(u) = K(\eta, u) \frac{(1 + |Du|^2)^{\frac{3}{2}}}{1 + u_t^2}, \quad \eta \in \Omega$$

is the *prescribed Levi curvature equation*
   The characteristic form of $\mathscr{L}_0$:

$$q_{\mathscr{L}_0}(\xi) := \xi_1^2 + \xi_2^2 + 2a\, \xi_1\, \xi_3 + 2b\, \xi_2\, \xi_3 + (a^2 + b^2)\, \xi_3^2$$

is the quadratic form related to the matrix

$$A := \begin{pmatrix} 1 & 0 & a \\ 0 & 1 & b \\ a & b & a^2 + b^2 \end{pmatrix}$$

whose eigenvalues are $\lambda_1 = 1$, $\lambda_2 = 1 + a^2 + b^2$, $\lambda_3 = 0$.

- Then $\mathscr{L}_0$ is a quasilinear, degenerate elliptic operator, which is *not elliptic at any point*.

Debiard and Gaveau [14] showed that $\mathscr{L}_0$ is *not variational*, i.e. cannot be written in divergence form.

3.2   The Dirichlet Problem for the Prescribed Levi Curvature Equation

Motivated by the following problems:

- construction of real surfaces of $\mathbb{C}^2$ with given boundary and prescribed Levi curvature
- construction of the envelopes of holomorphy

we now study the Dirichlet problem for the Levi equation

$$\begin{cases} \mathscr{L}_0(u) = K(\eta, u)\frac{(1+|Du|^2)^{\frac{3}{2}}}{1+u_t^2}, \ \eta \in \Omega \\ u = \varphi \qquad\qquad\qquad\qquad \text{on } \partial\Omega \end{cases}$$

This problem has received increasing attention since the early 1980s. The first existence results were obtained in the case $K \equiv 0$, with the analytic disc approach, by Bedford-Gaveau [3] and Bedford-Klingenberg [4]. However, to the best of our knowledge, their techniques do not work if $K \neq 0$.

3.3 The Prescribed Non Vanishing Levi Curvature Equations

In [39] Slodkowski and Tomassini introduced the use of PDE's techniques in studying

$$\begin{cases} \mathscr{L}_0(u) = K(\eta, u)\frac{(1+|Du|^2)^{\frac{3}{2}}}{1+u_t^2}, \ \eta \in \Omega \subset \mathbb{R}^3 \\ u = \varphi \qquad\qquad\qquad\qquad \text{on } \partial\Omega \end{cases}$$

when $K \neq 0$. However *the $C^{1,\alpha}$ techniques usually used for quasilinear elliptic equations do not work for the Levi equations.*
Nevertheless, Slodkowski and Tomassini, were able to prove

- L^∞ a-priori estimates for the gradient of the solutions

Then, using a Comparison Principle for $\mathscr{L}_0$, they proved the existence of $u \in$ Lip($\overline{\Omega}$) solving the equation *in the weak viscosity sense* of Crandall-Lions, and s.t. $u = \varphi$ on $\partial\Omega$.

3.4 Regularity of the Viscosity Solutions: Classical Solvability of the Dirichlet Problem

In [8] Citti et al. proved that:
every Lipschitz-continuous viscosity solution to the K-prescribed Levi curvature equation is of *class C^∞* if $K \in C^\infty$ and $K > 0$.
Thus, [39] plus [8] give the smooth solvability of the Dirichlet problem

$$\begin{cases} \mathscr{L}_0(u) = K(\eta, u)\frac{(1+|Du|^2)^{\frac{3}{2}}}{1+u_t^2}, \ \eta \in \Omega \subset \mathbb{R}^3 \\ u = \varphi \qquad\qquad\qquad\qquad \text{on } \partial\Omega \end{cases}$$

if K is smooth and strictly positive.

This kind of *hypoellipticity* result comes from the *sub-Riemannian structure* underlying $\mathscr{L}_0$.

3.5 $\mathscr{L}_0$'s Sub-Riemannian Structure and the Smoothness Result

We identify X and Y with the vector fields (i.e. first order PDO)

$$X = \partial_x + a\,\partial_t \quad \text{and} \quad Y = \partial_y + b\,\partial_t$$

The following crucial identities hold

$$\mathscr{L}_0 u = (X^2 u + Y^2 u)(1 + u_t^2) \tag{29}$$

$$[X, Y] = -\frac{\mathscr{L}_0 u}{1 + u_t^2}\,\partial_t \tag{30}$$

Then, the K prescribed Levi curvature equation can be written as

$$X^2 u + Y^2 u = K(\eta, u)\frac{(1 + |Du|^2)^{\frac{3}{2}}}{(1 + u_t^2)^2}$$

Moreover

$$[X, Y] = -q\,\partial_t, \quad q := K(\eta, u)\frac{(1 + |Du|^2)^{\frac{3}{2}}}{(1 + u_t^2)^2}$$

Summing up: u solves the K-prescribed Levi-curvature equation if

$$X^2 u + Y^2 u = K(\eta, u)\frac{(1 + |Du|^2)^{\frac{3}{2}}}{(1 + u_t^2)^2}$$

and, *if $K \neq 0$ everywhere ($\Leftrightarrow q \neq 0$ everywhere)*,

$$X = \partial_x + a\,\partial_t \quad Y = \partial_y + b\,\partial_t, \quad [X, Y] = -q\,\partial_t$$

are linearly independent at any point. In geometric language:

$$\text{rank Lie}\{X, Y\}(\eta) = 3 \quad \text{at any point } \eta \in \Omega \ (\subset \mathbb{R}^3)$$

If the vector fields X and Y were linear and smooth, this condition would imply the hypoellipticity of $X^2 + Y^2$, i.e. the smoothness of the weak distributional solutions to $X^2u + Y^2u = f$, $f \in C^\infty$.

In [8] we proved that a similar regularity result holds for the nonlinear Levi vector fields:

Theorem 6 (G.Citti-E.Lanconelli-A.M. (2002)). *If $u \in Lip(\Omega)$ solves the K-prescribed Levi-curvature equation*

$$X^2u + Y^2u = K(\eta, u)\frac{(1 + |Du|^2)^{\frac{3}{2}}}{(1 + u_t^2)^2}, \quad in \ \Omega \subset \mathbb{R}^3,$$

in the weak viscosity sense, then $u \in C^\infty$ if $K \in C^\infty, K > 0$.

Then :

- The Dirichlet problem for the K-prescribed Levi curvature equation in an open set $\Omega \subset \mathbb{R}^3$ has a smooth solution if $K \in C^\infty, K > 0$
- Every Lipschitz-continuous domain of $\mathbb{C}^2$ whose boundary has smooth and strictly positive Levi-curvature, is actually smooth.

Very recently, jointly with C. Gutierrez and E. Lanconelli in [18] we discovered that these results have *no counterpart in higher dimension.*

In next subsections we exploit the steps and the history of the proof of Theorem 6.

3.6   Strong Solutions

By differentiation of the equation and integration by parts, in [9] we obtained the following first regularity achievement

Theorem 7 (Citti, M. 1999). *If $K \in Lip(\Omega \times \mathbb{R})$, every viscosity solution $u \in Lip(\Omega)$ of the prescribed Levi curvature equation*

$$X^2u + Y^2u = K(\eta, u)\frac{(1 + |Du|^2)^{3/2}}{(1 + (\partial_t u)^2)^2}$$

is a strong solution, i.e. $Xu, Yu \in H^1_{loc}(\Omega)$ and u pointwise solves the equation a.e.

3.7   Sketch of the Proof of Theorem 7

In an open set Ω we fix a solution u of the regularized equation

$$\mathscr{L}_\varepsilon u = K(\eta, u)\frac{(1 + |Du|^2)^{3/2}}{(1 + (\partial_t u)^2)^2}$$

where $\mathscr{L}_\varepsilon$ denotes the elliptic operator (with minimum eigenvalue ε^2)

$$\mathscr{L}_\varepsilon u := \mathscr{L}u + \varepsilon^2 \frac{u_{tt}}{1 + u_t^2}$$

and

$$\mathscr{L}u := X^2 u + Y^2 u.$$

The uniform limit as ε goes to 0 of a sequence of solutions of the regularized equation is a viscosity solution of the Levi equation in the weak sense of Crandall-Ishii-Lions.

3.7.1 Linear Operator

In the sequel we shall denote by L_ε a linear elliptic operator formally defined as $\mathscr{L}_\varepsilon$:

$$L_\varepsilon = X^2 + Y^2 + T_\varepsilon^2,$$

where $T_\varepsilon = \varepsilon(1 + u_t^2)^{-1/2}\partial_t$, and the coefficients of the vector fields X and Y depend on the fixed solution u. Then we prove that the coefficients a and b of the vector fields and the two functions

$$\omega = \partial_t u, \quad and \quad v = arctg(u_t)$$

are solutions of

$$L_\varepsilon z = f,$$

with different functions f.

The proof of the regularity Theorem 7 is based on the regularity of the solutions of this linear equation in some Sobolev Spaces $W_\varepsilon^{1,2}$ naturally defined in terms of the vector fields X, Y and T_ε.

3.7.2 Caccioppoli-Type Estimates

Lemma 4. *For every $\phi \in C_0^\infty(\Omega)$ and $\delta \in (0, 1)$ there is $C_\delta > 0$ s.t.*

$$\int_\Omega (|Xz|^2 + |Yz|^2 + |T_\varepsilon z|^2)\phi^2 \leq \delta \int_\Omega (f^2 + |\partial_t a|^2 + |\partial_t b|^2 + |T_\varepsilon v|^2)\phi^2$$

$$+C_\delta \int_\Omega (|X\phi|^2 + |Y\phi|^2 + |T_\varepsilon \phi|^2 + (1 + \omega^2)\phi^2)z^2$$

Proof. Integrate by parts and recall that

$$X^*z = -Xz - \partial_t a \cdot z, \, Y^*z = -Yz - \partial_t b \cdot z, \, T_\varepsilon^* z = -T_\varepsilon z + \omega T_\varepsilon v \cdot z$$

Then use Cauchy Schwarz inequality   □

3.7.3 Elegant Equalities

With the previous notation, we have

Lemma 5.

$$\partial_t a = Yv - \omega Xv, \quad \partial_t b = -Xv - \omega Yv.$$

Proof. Since $a = Yu, b = -Xu$ (because we are computing tangential vector fields on the graph u) we have

$$\partial_t a = \partial_t Y u = [\partial_t, Y]u + Y\partial_t u = \omega \partial_t b + Y\omega,$$

$$\partial_t b = -\partial_t X u = -[\partial_t, X]u + X\partial_t u = -\omega \partial_t a - X\omega$$

$$\partial_t a = = \frac{-\omega X\omega + Y\omega}{1 + \omega^2} = -\omega Xv + Yv,$$

$$\partial_t b = = \frac{-\omega Y\omega - X\omega}{1 + \omega^2} = -\omega Yv - Xv \qquad\qquad □$$

3.7.4 Caccioppoli-Type Estimates

Lemma 6. *For every* $\phi \in C_0^\infty(\Omega)$

$$\int_\Omega (|Xa|^2 + |Ya|^2 + |T_\varepsilon a|^2)\phi^2 + \int_\Omega (|Xb|^2 + |Yb|^2 + |T_\varepsilon b|^2)\phi^2$$

$$\int_\Omega (|Xv|^2 + |Yv|^2 + |T_\varepsilon v|^2)\phi^2 \le C_1 \int_\Omega (|\nabla K|^2 + K^2)\phi^2$$

$$+C_2 \int_\Omega (|X\phi|^2 + |Y\phi|^2 + |T_\varepsilon \phi|^2 + (1 + \omega^2)\phi^2)(a^2 + b^2 + v^2)$$

In particular $\|\nabla a\|_{L^2_{loc}}, \|\nabla b\|_{L^2_{loc}}$ are uniformly bounded in ε.

3.8 Levi Flat

In [10] we proved the following Frobenius type result for strong solutions of the Levi flat equation.

Theorem 8 (Citti, M. 2001). *Every strong solution u of the Levi equation with $K \equiv 0$, is foliated by analytic curves and u is analytic on every leaf.*

The key Lemma in the proof is the following

Lemma 7. *Since $X^2u + Y^2u = 0$ we have $[X, Y] = 0$ and $X^2v + Y^2v = 0$*

Proof. We formally differentiate the equation $X^2u + Y^2u = 0$ with respect to ∂_t and we get

$$
\begin{aligned}
0 =&\, \partial_t(X^2u + Y^2u) \\
=&\, [\partial_t, X]Xu + X[\partial_t, X]u + X^2\omega + [\partial_t, Y]Yu + Y[\partial_t, Y]u + Y^2\omega \\
=&\, \partial_t a \partial_t Xu + X(\partial_t a\omega) + X^2\omega + \partial_t b \partial_t Yu + Y(\partial_t b\omega) + Y^2\omega \\
=&\, -\partial_t a \partial_t b + X(Yv\omega - Xv\omega^2) + X^2\omega \\
&\, + \partial_t b \partial_t a + Y(-Xv\omega - Yv\omega^2) + Y^2\omega \\
=&\, -(X^2v + Y^2v)\omega^2 - 2\omega(XvX\omega + YvY\omega) + X^2\omega + Y^2\omega \\
=&\, -(X^2v + Y^2v)\omega^2 - 2\omega(XvX\omega + YvY\omega) \\
&\, + X(Xv(1 + \omega^2)) + Y(Yv(1 + \omega^2)) = X^2v + Y^2v \qquad \square
\end{aligned}
$$

Let $D = (X, Y, T_\varepsilon)$. By Caccioppoli-type estimates we get that for every multi-index I and for every p the norms
$\|D^I u\|_{L^p_{loc}}, \|D^I v\|_{L^p_{loc}}$ are uniformly bounded in ε.
In particular the coefficients a, b of the vector fields X, Y are Lipschitz continuous and we can conclude by Frobenius theorem.

3.9 Classical Solution

If $K \neq 0$ at every point, in a joint work with Citti and Lanconelli [8] we obtained the following regularity result.

Theorem 9 (Citti, Lanconelli, M. 2002). *Every strong solution u of the Levi equation, with $K \neq 0$, $K \in C^\infty(\Omega)$, is smooth.*

3.10   Sketch of the Proof of Theorem 9

Since the minimum eigenvalue of

$$
A = \begin{pmatrix} 1 & 0 & a \\ 0 & 1 & b \\ a & b & a^2 + b^2 \end{pmatrix}
$$

is equal to zero for every $p \in \mathbb{R}^3$, the operator $\mathscr{L}_0$ in (29) is not elliptic at any point and the regularity results for viscosity solutions of non-linear elliptic and parabolic equations cannot be applied to our case. We have to introduce a completely different procedure, based on the particular structure of the Levi equation.

3.10.1   Lie Bracket

Let us recall that in this situation the Lie-bracket of the first order differential operators X and Y is

$$
[X, Y] = -\frac{\mathscr{L}_0 u}{1 + u_t^2} \partial_t.
$$

3.10.2   Interpolation Inequalities

The classical elliptic regularization procedure is based on Sobolev inequalities and on a priori estimates of Caccioppoli type. In the present situation neither the classical Caccioppoli inequality holds, since the vector fields are not selfadjoint, nor the Sobolev inequality, since the coefficients of the vector fields are only bounded.

To overcome these difficulties we first prove an interpolation inequality, which will play a role similar to the Sobolev one in the classical setting.

3.10.3   Interpolation Inequalities

Lemma 8. *Let C be a positive constant such that*

$$
||a||_\infty + ||b||_\infty + ||v||_\infty \le C,
$$

then for every function $z \in C^\infty$, $\phi \in C_0^\infty$, we have

$$
\int |Xz|^3 \phi^6 \le c \int |D_\varepsilon (Xz)|^2 \phi^6 + c \int \left(|D_\varepsilon \phi|^6 + \phi^6 \right)(1 + z^6)
$$

where D_ε is the vector (X, Y, T_ε) and the constant $c > 0$ only depends on C and K. An analogous inequality is also satisfied if we replace Xz with Yz or $T_\varepsilon z$.

3.10.4 Linear Operator

- We first establish some a priori estimates in the intrinsic directions X and Y, weaker than the classical Caccioppoli one.
- Using these inequalities together with the interpolation ones, we prove a priori estimates in $W_\varepsilon^{m,p}$, for solutions z, which hold under very general assumptions on the commutators of the vector fields, but requires some strong a priori estimates on the derivative $\partial_t z$.
- To estimate the derivative ∂_t, we recall that it can be expressed in term of the commutator of the vector fields.

3.10.5 Non Linear Equation

- We then use in an essential way the nonlinearity of the equation: we apply the interpolation and Caccioppoli inequalities to the derivative $\partial_t v$ to obtain a L^2 estimate for $X v_t$ and $Y v_t$.
- Since $v_t = \frac{u_{tt}}{1+u_t^2}$, then v_t has to be considered a derivative of weight 4 of u, while $X v_t$ and $Y v_t$ are derivatives of weight 5 of the same function. Once proved the summability of these derivatives with respect to t, we obtain analogous estimates for any derivation of weight 5 and 4.

3.10.6 Estimate of $\partial_t v$

Lemma 9. *If $K \neq 0$ then for every $\phi \in C_0^\infty$*

$$\int (|\partial_t v|^3 + |D_\varepsilon \partial_t v|^2)\phi^6 \leq C \int (|\partial_t \phi|^2 + |D_\varepsilon \partial_t \phi|^2)(v^2 + |D_\varepsilon v|^2)$$

Proof. Write $\partial_t = -\frac{1}{q}[X,Y]$, integrate by parts and use Cauchy Schwarz   □

In particular the coefficients a and b of the vector fields are now regular, and we can apply a Sobolev type inequality.

3.10.7 Non Linear Equation

By applying the previous results to the non linear equation we get that

- the derivatives of weight 5 belong to L^2
- the derivatives of weight 4 belong to L^4, because the homogeneous dimension in this case is $Q = 4$ and $\frac{1}{2} - \frac{1}{Q} = \frac{1}{4}$
- the derivatives of weight 3 belong to L^p for every p,

- the derivatives of weight 2 belong to suitable classes C^α for every $\alpha \in]0, 1[$ and we will write $u \in C_X^{2,\alpha}$

Now, we can freeze the coefficients $a, b \in C_X^{1,\alpha}$ of the vector fields.

3.10.8 Freezing Method

Let $\eta_0 = (x_0, y_0, t_0) \in \Omega$. Given $a \in C_X^1$ we define the first order Taylor polynomial of a in the horizontal directions as

$$P_{\eta_0} a(\eta) = a(\eta_0) + Xa(\eta_0)(x - x_0) + Ya(\eta_0)(y - y_0), \quad \eta = (x, y, t).$$

Now we define the frozen vector fields

$$X_{\eta_0} = \partial_x + P_{\eta_0} a(\eta)\partial_t, \quad Y_{\eta_0} = \partial_y + P_{\eta_0} b(\eta)\partial_t.$$

We have

$$[X_{\eta_0}, Y_{\eta_0}] = (Xb - Ya)(\eta_0)\partial_t$$

and $(Xb - Ya)(\eta_0) = -(X^2 u + Y^2 u)(\eta_0) = -q(\eta_0) \neq 0$.

Then the vector fields X_{η_0}, Y_{η_0} satisfy Hörmander condition and the linear operator $L_{\eta_0} = X_{\eta_0}^2 + Y_{\eta_0}^2$ is hypoelliptic.

By using integral formulas for u in terms of the fundamental solution of L_{η_0} we deduce that $u \in C^{2,\alpha}$. By a bootstrap argument we then conclude that $u \in C^\infty$.

In [11] a freezing method with second order Taylor polynomials has been introduced to prove the following regularity theorem.

Theorem 10 (Citti, M. 2002). *If the prescribed curvature K is smooth and it has some first order zeros, then classical solutions of the prescribed Levi curvature equation are C^∞.*

4 Levi Curvatures in Higher Dimension

In this section we shall explicitly compute the Levi curvatures for a real hypersurface in $\mathbb{C}^{n+1}$. We then shall study the Dirichlet problem for the prescribed Gauss-Levi curvature equation in the class of Lipschitz continuous viscosity solutions, under very natural conditions on the prescribed curvature K and on the boundary data. To state our result we need a suitable notion of a generalized Levi pseudoconvex solution to the prescribed Gauss Levi curvature equation, which is a fully non linear and a non elliptic pde. This is provided by the notion of viscosity solution in the sense of Crandall et al. [12]. The modifications needed to adapt the theory

of viscosity to equations of Levi Monge Ampère have been studied in the paper
[13] and we will recall them for reader convenience. We will also use a comparison
principle for viscosity solution, Theorem 14, and existence results of continuous and
of Lipschitz continuous viscosity solution of the prescribed Gauss Levi curvature
equation, Theorem 15, which have been proved in [13]. The problem of classical
regularity of Lipschitz continuous viscosity solutions is widely still open. However,
we shall show that the fully nonlinear PDE of the prescribed Gauss-Levi curvature
equation has the following hypoellipticity property: every classical solution is
smooth, if $K \neq 0$ at any point and $K \in C^{\infty}$.

4.1 Normalized Levi Form

If M is represented in a neighborhood of p by $f = 0$, $\nabla f \neq 0$ and the orientation
of M is so that $\nabla f / |\nabla f|$ is the outward pointing normal, then the *normalized Levi
form* at p is the Hermitian form on $H_p^{1,0} M \times H_p^{0,1} M$

$$\mathcal{L}_p(Z_j, \bar{Z}_k) = \langle \frac{\bar{\partial}\partial f(p)}{|\partial f|} Z_j, \bar{Z}_k \rangle,$$

$\{Z_1, \ldots, Z_n\}$ orthonormal basis of $H_p^{1,0} M$.

Let us recall that $\mathcal{L}_p$ contains less information on the shape of M than $\mathbb{II}_p$
because

$$\mathcal{L}_p(Z, \bar{Z}) = \frac{\mathbb{II}_p(X, X) + \mathbb{II}_p(Y, Y)}{2}, \quad \forall Z = \frac{X - iY}{\sqrt{2}} \in H_p^{1,0} M$$

Remark 5. In this situation

- $M = \partial D = \{z \in \mathbb{C}^{n+1} : f(z) = 0\}$ is a real manifold of $\mathbb{C}^{n+1} \equiv \mathbb{R}^{2n+2}$ of dim $2n + 1$
- $H_p^{1,0}(\partial D) = \{h \in \mathbb{C}^{n+1} : \langle h, \partial_p f \rangle = 0\}$ and
        dim $H_p^{1,0}(\partial D) = n$, as complex vector space
        dim $H_p^{1,0}(\partial D) = 2n$, as real vector space
- in passing from ∂D to $H_p^{1,0}(\partial D)$ we lose a real dimension

4.2 Gauss-Levi Curvature for a Real Hypersurface in $\mathbb{C}^{n+1}$

Definition 19. A domain D is *pseudoconvex* if the Levi form of its boundary is
positive semidefinite at every point of ∂D.

The Gauss Levi Curvature of ∂D in $p \in \partial D$ is the real number

$$K_{\partial D}(p) = \prod_{j=1}^{n} \lambda_j,$$

where $\lambda_j =$ are the eigenvalues of the normalized Levi form.

By a linear algebra argument we can write it as in [31]

$$K_{\partial D}(p) = \frac{1}{|\partial_p f|^{n+2}} LMA(f), \quad LMA(f) := -\det \begin{pmatrix} 0 & f_{\bar z} \\ f_z & f_{z,\bar z} \end{pmatrix}. \qquad (31)$$

The operator *LMA* is called the Levi Monge-Ampère operator.

4.3   m-Pseudoconvexity

Definition 20. A domain D of $\mathbb{C}^{n+1}$ is (strictly) *m-pseudoconvex* if, for every $j \in \{1,\ldots,m\}$,

$$\sigma^{(j)}(\lambda_1,\ldots,\lambda_n) := \sum_{1 \le i_1 < \cdots < i_j \le n} \lambda_{i_1} \cdots \lambda_{i_j} \ge 0 \, (> 0)$$

at every point $p \in \partial D$. Here and in the sequel $\sigma^{(j)}$ is the j-th elementary symmetric function.

For every $m \in \{1,\ldots,n\}$ we define *m-Levi curvature* of ∂D

$$K_{\partial D}^{(m)}(p) = \sigma^{(m)}(\lambda) / \binom{n}{m}.$$

Precisely,

- $K^{(n)}$ is the Gauss- Levi curvature,
- $K^{(1)}$ is the mean Levi curvature

$$K_{\partial D}^{(1)} = \frac{\lambda_1 + \cdots + \lambda_n}{n}.$$

4.4   General Expressions for $K^{(m)}$

By linear algebra arguments in [31] we recognized that if $m = n$ the Gauss-Levi curvature is

$$K_{\partial D}^{(n)}(p) = -\frac{1}{|\partial_p f|^{n+2}} \det \begin{pmatrix} 0 & f_{\bar{1}} & \cdots & f_{n\bar{+}1} \\ f_1 & & & \\ \vdots & & f_{j,\bar{k}} & \\ f_n & & & \end{pmatrix}$$

and if $1 \leq m \leq n$

$$K_{\partial D}^{(m)}(p) = -\frac{1}{\binom{n}{m}} \frac{1}{|\partial_p f|^{m+2}} \sum_{1 \leq i_1 < \cdots < i_{m+1} \leq n+1} \Delta_{i_1, \ldots, i_{m+1}}(f)$$

where $\quad \Delta_{i_1, \ldots, i_{m+1}}(f) := \det \begin{pmatrix} 0 & f_{\bar{i}_1} & \cdots & f_{\overline{i_{m+1}}} \\ f_{i_1} & & & \\ \vdots & & f_{i_j, \bar{i}_k} & \\ f_{i_{m+1}} & & & \end{pmatrix}.$

4.5 Levi Curvatures in $\mathbb{C}^{n+1}, n \geq 1$: Geometric Contents and Properties

Even if the Levi curvatures contain less information about the shape of the domain, they have the following geometric contents

- (Isoperimetric inequality) If $D \subset \mathbb{C}^{n+1}$ is bounded and smooth, $K_p^{(m)}(\partial D) \geq 0$ for all $p \in \partial D$, then

$$\int_{\partial D} \left(\frac{1}{K_p^{(m)}(\partial D)} \right)^{\frac{1}{m}} d\sigma(p) \geq (2n+2)\text{meas}(D)$$

  where $\text{meas}(D)$ is the Lebesgue measure of D (see [28]).
- A biholomorphic change of variables F preserves the Levi curvatures of every smooth domain of $\mathbb{C}^{n+1}$, if and only if F is a linear unitary map (see [21]).
- The Levi curvatures are preserved by complex reflections.

4.6 Domains of Holomorphy and Levi Curvatures

Summing up, we can characterize domains of holomorphy by studying the sign of Levi curvatures of the boundary. Precisely, we have: D is a domain of holomorphy iff $K_{\partial D}^{(m)}(p) \geq 0$ for $m = 1, \cdots, n$ and $\forall p \in \partial D$.

Indeed, the following statements are equivalent

- D is a domain of holomorphy
- D is pseudoconvex
- The Levi form is nonnegative definite at any point of ∂D
- The eigenvalues of the normalized Levi form are nonnegative
- $K_{\partial D}^{(m)}(p) \geq 0$ for $1 \leq m \leq n$ and $\forall p \in \partial D$.

4.7   Levi Curvatures of Graphs: Levi Curvature Operators

Let $\Omega \subseteq \mathbb{R}^{2n+1}$ be open and $u : \Omega \to \mathbb{R}$, u of class C^2.

We identify: $\mathbb{R}^{2n+1} \times \mathbb{R}$ with $\mathbb{C}^{n+1}$, and denote
$$z = (x, y) \, z_j = x_{2j-1} + i x_{2j}, \, 1 \leq j \leq n, \quad z_{n+1} = x_{2n+1} + iy$$

We let

$$\gamma(u) = \{(x, y) \in \Omega \times \mathbb{R} : y = u(x)\} \equiv \text{ graph of } u$$

$\gamma(u)$ is part of the boundary of

$$\Gamma(u) = \{(x, y) \in \Omega \times \mathbb{R} : y > u(x)\} \subset \mathbb{C}^{n+1}.$$

We will say that u is (strictly) m-pseudoconvex if $\Gamma(u)$ is (strictly) m-pseudoconvex.
For $1 \leq m \leq n$ we define the m-th Levi curvature operator as

$$\mathscr{L}^{(m)}(u)(x) := K_p^{(m)}(\gamma(u)), \quad x \in \Omega, \quad p = (x, u(x))$$

4.7.1   Structure of the Levi Curvature PDO's

The operator $\mathscr{L}^{(m)}$ can be written as follows

$$\mathscr{L}^{(m)}(u)(x) = \sum_{j,k=1}^{2n} a_{j,k} Z_j Z_k(u), \quad u \in C^2(\Omega, \mathbb{R}), \quad \Omega \subseteq \mathbb{R}^{2n+1}.$$

where

$$Z_j = \partial_{x_j} + a_j \partial_{x_{2n+1}}, \quad a_j = a_j(Du), \quad j = 1, \cdots, 2n$$

and $(a_{j,k})_{k,j=1,\dots,2n}$ is symmetric, $a_{j,k} = a_{j,k}(Du, D^2 u)$

Then:

$$Z_{j,k} := [Z_j, Z_k] = q_{j,k} \, \partial_{x_{2n+1}}$$

and we have the following *subellipticity properties* of $\mathscr{L}^{(m)}$: if u is m-strictly pseudoconvex:

- $\forall x \in \Omega, \ q_{j,k}(x) \neq 0$ for suitable j,k
- $(a_{j,k})_{k,j=1,\dots,2n}$ is strictly positive definite.

4.7.2 Some Comments

If u is m-strictly pseudoconvex:

- $\mathscr{L}^{(m)}$ is elliptic only along the $2n$ linearly independent directions

$$Z_j = \partial_{x_j} + a_j \partial_{x_{2n+1}} \equiv e_j + a_j e_{2n+1}, j = 1, \dots, 2n$$

- The **missing ellipticity direction** e_{2n+1} is recovered by commutation:

$$Z_{j,k} := [Z_j, Z_k] = q_{j,k} \partial_{x_{2n+1}} \equiv q_{j,k} e_{2n+1}$$

- The commutation property can be restated as follows

$$\dim(\text{span}\{Z_j(x), Z_{j,k}(x) \ : \ j,k = 1, \dots, 2n\}) = 2n + 1, \quad \forall x \in \Omega$$

- $\mathscr{L}^{(m)}$ is a PDO in $\mathbb{R}^{2n+1}$, which is fully nonlinear if $n > 1$

4.7.3 Some Results

From the subelliptic properties of $\mathscr{L}^{(m)}$ one gets:

Theorem 11 (STRONG COMPARISON PRINCIPLE). *Let* $u, v : \Omega \to \mathbb{R}, \Omega \subseteq \mathbb{R}^{2n+1}$ *open and connected.*
Assume u and v strictly m-pseudoconvex and

(i) $u \leq v$ *in* $\Omega, u(x_0) = v(x_0)$ *at* $x_0 \in \Omega$
(ii) $\mathscr{L}^{(m)}(u) \geq \mathscr{L}^{(m)}(v)$ *in* Ω

Then $u = v$ *in* Ω.

The strong comparison principle has been proved in [6] for $n = 1$ and in [31] for the general case.

Theorem 12 (SMOOTHNESS OF $C^{2,\alpha}$-SOLUTIONS). *Let $u \in C^{2,\alpha}(\Omega)$ be a strictly m-pseudoconvex solution to the K-prescribed Levi curvature equation*

$$\mathcal{L}^{(m)}(u) = K(\cdot, u, Du) \quad in \ \Omega$$

If K is strictly positive and C^{∞} in its domain, then $u \in C^{\infty}(\Omega)$

The smoothness of $C^{2,\alpha}$-solutions has been proved in [7] for $n = m = 1$, in [32] for $n = m \geq 1$ and in [30] for the general case $1 \leq m \leq n$.

4.8 Hypoellipticity

Precisely, in [32] we proved

Theorem 13 (Lascialfari, M. (2004)). *If $u \in C^{2,\alpha}(\Omega)$ is a pseudoconvex solution of*

$$LMA(u) = q(\cdot, u, Du)$$

where $\Omega \subset \mathbb{R}^{2n+1}$ and $q \in C^{\infty}(\Omega \times \mathbb{R} \times \mathbb{R}^{2n+1})$, $q > 0$, then $u \in C^{\infty}(\Omega)$. Here $LMA(u) = LMa(u-y)$ is the Levi Monge-Ampère operator in (31) for the graph u.

4.9 Sketch of the Proof of Theorem 13

If u is a strictly Levi pseudoconvex solution,

$$LMA(u) = \sum_{i,j=1}^{2n} a_{ij}(x) Z_i Z_j u,$$

where $Z_j = \partial_j + a_j(x)\partial_{x_{2n+1}}$, $a_j(x) = a_j(Du(x))$ and $a_{ij}(x) = a_{ij}(Z^2 u) = a_{ij}(Du(x), D^2 u(x))$.

For all $K \subset\subset \Omega$ there exists $C > 0$ such that

$$\sum_{i,j=1}^{2n} a_{ij}(x)\xi_i\xi_j \geq C \sum_{j=1}^{2n} \xi_j^2, \quad \forall \xi \in \mathbb{R}^{2n}, \quad \forall x \in K.$$

The *LMA* operator is *elliptic in 2n linearly independent directions* and we have the lack of ellipticity in one direction.

4.9.1 Frozen Operator

If u is a fixed C^2 strictly Levi pseudoconvex solution and $x_0 \in \Omega$, we define the linear operator

$$L_{x_0} = \sum_{i,j=1}^{2n} a_{ij}(x_0) Z_{x_0,i} Z_{x_0,j},$$

where, with the notation of subsection 3.10.8, $Z_{x_0,j} = \partial_j + P_{x_0} a_j(x) \partial_{x_{2n+1}}$, $P_{x_0} a_j(x) = a_j(x_0) + \sum_{i=1}^{2n} Z_i a_j(x_0)(x_i - x_{0,i})$ and $a_{ij}(x_0) = a_{ij}(Z^2 u(x_0))$.

The linear second order operator L_{x_0} is *elliptic in 2n linearly independent directions* and we have the lack of ellipticity in one direction.

4.9.2 Brackets

A direct computation shows that

$$[Z_{x_0,i} Z_{x_0,j}] = q_{ij}(x_0) \partial_{x_{2n+1}}.$$

Since u is a strictly pseudoconvex solution we get

- there is (i, j) such that $q_{ij}(x_0) \neq 0$. Hence

  $$dim \left(Span \left\{ Z_{x_0,i}, [Z_{x_0,i}, Z_{x_0,j}], \; i, j = 1, \ldots, 2n \right\} \right) = 2n + 1$$

  at every point Ω.
- The linear operator L_{x_0} is hypoelliptic.
- By using integral formulas for $u \in C^{2,\alpha}$ in terms of the fundamental solution of L_{x_0} we deduce that $u \in C_{loc}^{3,\beta}$, for every $0 < \beta < \alpha$. By a bootstrap argument we can conclude that $u \in C^\infty$.

4.10 A More Explicit Form of the Gauss-Levi Curvature Equation

In order to deal with viscosity solutions we shall write the Gauss-Levi curvature equation in a more explicit form as a degenerate elliptic fully nonlinear PDE.

A function u solves the K-prescribed Gauss-Levi curvature equation if

$$\mathcal{L}(K; u) := LMA(u) - K(x, u) F(Du) = 0, \quad LMA(u) = \det A(Du, D^2 u)$$

Du = gradient of u, $D^2 u$ = Hessian of u with respect to all the variables $x \in \mathbb{R}^{2n+1}$,

- $F(Du) = 2^n \frac{(1+|Du|^2)^{\frac{n+2}{2}}}{1+(\partial_{x_{2n+1}}u)^2}$
- $A(Du, D^2u) := \Sigma \, D^2u \, \bar{\Sigma}^T$ is an $n \times n$ matrix with complex entries
- $\Sigma = (I_n, -iI_n, a - ib)$, a and b are vectors with components

$$a_\ell := a_\ell(Du) = \frac{\partial_{x_{2\ell}}u - \partial_{x_{2\ell-1}}u \, \partial_{x_{2n+1}}u}{1 + (\partial_{x_{2n+1}}u)^2}$$

$$b_\ell := b_\ell(Du) = \frac{-\partial_{x_{2\ell-1}}u - \partial_{x_{2\ell}}u \, \partial_{x_{2n+1}}u}{1 + (\partial_{x_{2n+1}}u)^2},$$

4.11 A More Explicit Form of Pseudoconvexity

Remark 6. u is pseudoconvex iff $A(Du, D^2u) \geq 0$.

Remark 7. If u is pseudoconvex then $LMA(u) = \det A(Du, D^2u)$ is a degenerate elliptic operator.

4.12 The Dirichlet Problem for LMA

Viscosity techniques work very well to study the Dirichlet problem

$$\begin{cases} \det A(Du, D^2u) = K(x, u) \, F(Du) & x \in \Omega \subset \mathbb{R}^{2n+1} \\ u = \phi & \text{on } \partial\Omega \end{cases} \tag{32}$$

Under quite natural assumptions, this problem has *a Lipschitz continuous viscosity solution* (see [13, 40]).

4.13 Viscosity Solutions to the Prescribed Levi Curvature Equation

By following [13] we give the following definitions

Definition 21. Given u and φ real functions in Ω, we say that φ *touches* u from below (above) at $x_0 \in \Omega$ if $u(x_0) = \varphi(x_0)$ and $\varphi(x) \leq (\geq)u(x)$ for all x in some neighborhood of x_0.

Definition 22. $u : \Omega \to \mathbb{R}$ is a *viscosity subsolution* if u is in $BUSC(\Omega)$, and for every $\varphi \in C^2(\Omega)$ and for every $x_0 \in \Omega$ such that φ touches u from above at x_0, we have $A(D\varphi, D^2\varphi)(x_0) \geq 0$ and $\mathcal{L}(K, \varphi)(x_0) \geq 0$.

Definition 23. $u : \Omega \to \mathbb{R}$ is a *viscosity supersolution* if u is in $BLSC(\Omega)$, and for every $\varphi \in C^2(\Omega)$ and for every $x_0 \in \Omega$ such that φ touches u from below at x_0 and $A(D\varphi, D^2\varphi)(x_0) \geq 0$ we have $\mathcal{L}(K, \varphi)(x_0) \leq 0$.

Definition 24. A function $u : \Omega \to \mathbb{R}$ is a *viscosity solution* if it is both a viscosity subsolution and a viscosity supersolution.

4.13.1 Hartogs's Pseudoconvexity

Remark 8. The epigraph of a viscosity solution u is a pseudoconvex domain of $\mathbb{C}^{n+1}$ in the Hartogs's generalized sense of Definition 16. The proof of this statement is implicitly contained in [19, Theorem 4.1.27]) and we leave details to the reader.

We express this property by simply saying that u is pseudoconvex.

4.14 Hypotheses on K

In order to study the Dirichlet problem (32) we assume that the prescribed function $K : \overline{\Omega} \times \mathbb{R} \to [0, +\infty)$ is bounded and continuous and it satisfies the following hypotheses

(H1) STRICTLY MONOTONICITY
  for every $R > 0$, there is $\ell_R > 0$, such that, for all $x \in \overline{\Omega}$, and $-R \leq v \leq u \leq R$ $\ell_R(u - v) \leq K(\cdot, u) - K(\cdot, v)$
(H2) MODULUS OF CONTINUITY
  for every $R > 0$, there is ω_R such that $\omega_R(s) \to 0$ for $s \to 0^+$ and $|K(x, u) - K(y, u)| \leq \omega_R(|x - y|)$ for every $(x, y) \in \overline{\Omega}$ and $|u| \leq R$.
(H3) BOUNDARY CONDITION
  $\Omega \times i\mathbb{R}$ is strictly pseudoconvex and its curvature $K_{\partial\Omega \times i\mathbb{R}}$ satisfies $\sup_{\overline{\Omega} \times \mathbb{R}} K < K_{\partial\Omega \times i\mathbb{R}}(x_0)$ for every $x_0 \in \partial\Omega$

4.15 Comparison Principle

Theorem 14 (COMPARISON PRINCIPLE). *Assume* (H1)–(H3), u *sub- and* v *supersolution of* (32), *then*

$$u \leq v \quad in \; \bar{\Omega}.$$

Proof. Use Ishii method. We refer to [13] for the adaptation of this method to the Levi Monge-Ampère equation   □

4.16 Existence and Uniqueness

Theorem 15 (UNIQUENESS AND EXISTENCE THEOREMS). *If* (H1)–(H3) *hold,
then* $\forall \phi \in C(\partial\Omega)$ $\exists! u$ *viscosity solution of* (32).
Moreover, if $\phi \in Lip(\partial\Omega)$ *and* $K \in Lip(\Omega \times \mathbb{R})$ *then* $u \in Lip(\Omega)$.

Proof. The ingredients of the proof in [13] are

- Perron method + comparison principle + existence of a particular subsolution and of a particular supersolution
- Interior gradient estimates + boundary gradient estimates
- To prove boundary gradient estimates, choose λ such that

$$\underline{u} = \phi - \lambda d \le u \le \phi + \lambda d = \overline{u}$$

near the boundary, where d is the boundary distance. $\qquad\square$

4.17 The Dirichlet Problem for LMA

Summing up:

- Viscosity techniques work to study the Dirichlet problem (32) and under quite natural assumptions, the problem has *a Lipschitz continuous viscosity solution* by [13].
- In dimension $n > 1$ the problem of the $C^{2,\alpha}$ *solvability* of this problem is still widely open.
- The fully nonlinear elliptic techniques do not work for the Levi equations
- In dimension $n > 1$ Lipschitz continuous viscosity solutions are not expected to be smooth (even if the data K is smooth)

5 A Negative Regularity Result for LMA

In this section we prove the existence of nonsmooth viscosity solutions for the Levi Monge Ampère equation for $n > 1$. This generalizes a result of Pogorelov for the Monge Ampère equation (see [17]).

In 1971 Pogorelov [34] showed that convex generalized solutions of the Monge Ampère equation

$$\det D^2 u = f(x) \tag{33}$$

in a domain $\Omega \subset \mathbb{R}^{n+1}$, $n > 1$, need not be of class C^2, even if f is positive and smooth. Urbas in [44] proved that this absence of classical regularity is not confined to equations of Monge Ampère type, but in fact occurs for the m-th elementary symmetric functions in the Hessian and the equation of prescribed m-th curvature, where in each case $m \geq 3$.

Recently, in [18] we showed that a similar result holds for the Levi Monge Ampère equation.

Let us remark the following facts:

- It is well known [8] that for $n = 1$, if the prescribed Levi curvature K is smooth and positive, then every Lipschitz continuous viscosity solution of the prescribed Levi curvature equation is smooth. Moreover, if K is strictly monotone increasing with respect to u, then by the interior gradient bound in [27] every continuous viscosity solution is locally Lipschitz continuous
- If the prescribed Gauss Levi curvature is smooth and positive, the smoothness of classical solutions is non trivial because the equation is non elliptic, and it was proved in [32]

The main result of this section is Theorem 16, where we show the existence of a viscosity solution of the Dirichlet problem in a small ball B_ε for the prescribed Gauss Levi curvature equation (34) with appropriate boundary data of class $C^{1,1-2/n}$ and by using suitable comparable functions of Pogorelov type to show that the solution cannot have better regularity than this in the interior. The low degree of regularity of the boundary data is necessary for our proof and by using our technique we cannot obtain (nor do we expect) a similar result for boundary data of class $C^{1,1-\alpha}$ for $\alpha > 1 - 2/n$.

5.1 Notation and Main Theorem

Let $\Omega \subset \mathbb{R}^{2n+1}$ be open. The function $u : \Omega \to \mathbb{R}$ is a classical solution to the K-prescribed Levi curvature equation if $u \in C^2(\Omega)$ and satisfies the pde

$$\mathscr{L}(K, u) := \det A(Du, D^2u) - K(\xi, u) F(Du) = 0, \qquad (34)$$

for all $\xi \in \Omega$, where Du and D^2u denote the gradient and Hessian of u in all the variables x, y, t, respectively,

$$F(Du) = 2^n \frac{(1 + |Du|^2)^{\frac{n+2}{2}}}{1 + (\partial_t u)^2},$$

and the matrix $A(Du, D^2u)$ is an $n \times n$ matrix defined as follows.

For $\ell = 1, \cdots, n$, let

$$a_\ell := a_\ell(Du) = \frac{\partial_{y_\ell} u - \partial_{x_\ell} u \, \partial_t u}{1 + (\partial_t u)^2}, \qquad b_\ell := b_\ell(Du) = \frac{-\partial_{x_\ell} u - \partial_{y_\ell} u \, \partial_t u}{1 + (\partial_t u)^2}, \qquad (35)$$

and let a be the column vector with components a_ℓ, and b the column vector with components b_ℓ. Let Σ be the $n \times (2n + 1)$ complex matrix defined by

$$\Sigma = (I_n, -i I_n, a - ib),$$

where I_n is the $n \times n$ identity matrix. Then the matrix $A(Du, D^2u)$ is defined by

$$A(Du, D^2u) := \Sigma \, D^2 u \, \bar{\Sigma}^T. \tag{36}$$

Equation (34) geometrically means that the hypersurface M_u, graph of the solution u, has Levi curvature equal K, agreeing to let

$$M_u := \{z \in \mathbb{C}^{n+1} \; : \; z = (x + iy, t + iu(x, y, t)), (x, y, t) \in \Omega\}.$$

Let B_r = Euclidean ball in $\mathbb{R}^{2n+1}$ centered at the origin with radius r and assume $K \in C^\infty(B_1 \times \mathbb{R})$ is strictly positive and $s \mapsto K(\cdot, s)$ is increasing.

The main result of this section is

Theorem 16. *There exists $r \in (0, 1)$ and a pseudoconvex function $u \in Lip(\bar{B}_r)$ solving*

$$\det A(Du, D^2u) = K(\xi, u) \, F(Du) \quad in \quad B_r,$$

in the weak viscosity sense, and such that

- $u \in C^\infty(B_r)$ *if $n = 1$*
- $u \notin C^1(B_r)$ *if $n = 2$,*
- $u \notin C^{1,\beta}$ *for any $\beta > 1 - \frac{2}{n}$, if $n \geq 2$.*

Remark 9. In every dimension $n \geq 1$: if u were $C^{2,\alpha}$ then u would be C^∞.

Remark 10. Comments on the case $n = 2$

- In the case $n = 2$ the classical Pogorelov counterexample for $\det D^2u = f$ does not hold, and you have a $C^{1,\alpha}$ regularity theory for convex viscosity solutions.
- In the case $n = 2$ for every K smooth and positive, we will build a Lipschitz continuous viscosity solutions of the LMA equation such that $u \notin C^1(B_r)$

In order to guarantee the existence of a viscosity solution on B_R we argue as follows

- We fix $0 < R < 1$ such that

$$\sup_{B_1 \times \mathbb{R}} K < \frac{1}{2 R^n} \tag{37}$$

• Geometrically, this condition means that

$$\sup_{B_1 \times \mathbb{R}} K < \inf_{q \in \partial C_R} K_{\partial C_R}^{(n)}(q), \tag{38}$$

where C_R is the cylinder of $\mathbb{C}^{n+1}$

$$C_R := \{(x + iy, t + i\tau) : (x, y, t) \in B_R, \tau \in \mathbb{R}\},$$

and $K_{\partial C_R}^{(n)}(q)$ denotes the Levi-curvature of the boundary of C_R at the point q. Indeed, since the real function

$$f(q) = |x|^2 + |y|^2 + t^2 - R^2, \quad q = (x + iy, t + i\tau),$$

is a defining function of C_R, we get

$$K_{\partial C_R}^{(n)}(q) = \frac{R^2 + t^2}{2 R^{n+2}}, \quad \forall \ q = (x + iy, t + i\tau) \in \partial C_R.$$

As a consequence: $\inf_{q \in \partial C_R} K_{\partial C_R}^{(n)}(q) = \frac{1}{2R^n}$, showing that (37) is equivalent to (38).

5.2   A Strict Subsolution

To prove the existence part of Theorem 16, we show the following lemma, which provides a strict subsolution to (34).

Lemma 10. *Let* $0 < r < R < 1$ *and let* $\varphi \in C^2(B_1)$ *be a convex function. For each* $\lambda > 0$ *define*

$$u_\lambda(\xi) := \varphi(\xi) - \lambda d(\xi), \quad \xi \in B_1,$$

where $d(\xi) := r^2 - |\xi|^2$. *Then, there exists* $\lambda^* > 0$, *only depending on* r *and* $\sup_{B_R} |D\varphi|$, *such that*

$$\mathscr{L}(K, u_\lambda) > 0 \text{ in } B_r, \quad \text{for every } \lambda > \lambda^*. \tag{39}$$

5.3   Existence Result

Lemma 11. *If* $0 < r < R < 1$ *and* $\varphi \in C^2(B_1)$ *is a convex function such that* $\mathscr{L}(K, \varphi) \leq 0$ *in* B_1, *then the Dirichlet problem*

$$\mathscr{L}(K, u) = 0, \text{ in } B_r, \qquad u = \varphi \text{ on } \partial B_r \tag{40}$$

has a viscosity solution $u \in Lip\,(\overline{B_r})$ satisfying

$$\|u\|_{L^\infty(\overline{B_r})} + \|u\|_{Lip\,(\overline{B_r})} \leq C \tag{41}$$

where C only depends on r, $\|\varphi\|_{L^\infty(\overline{B_R})}$, $\|D\varphi\|_{L^\infty(\overline{B_R})}$, and $\|DK\|_{L^\infty(\overline{B_R})}$.

Remark 11. C is independent of $\|D^2\varphi\|_{L^\infty}$

Proof. Let $u_\lambda = \varphi - \lambda d$ be the function given by the previous Lemma 10 with $\lambda > \lambda^*$. Then $u_\lambda \in C^2(\overline{B_r})$ and is a classical subsolution to $\mathscr{L}(K, u) = 0$ in B_r. Moreover, $u_\lambda = \varphi$ on ∂B_r.

On the other hand, since $\mathscr{L}(K, \varphi) \leq 0$ in B_1, φ is a classical supersolution to $\mathscr{L}(K, u) = 0$ in B_r.

Then, the Dirichlet problem has a viscosity solution $u \in C(\overline{B_r})$ verifying $u_\lambda \leq u \leq \varphi$ in $\overline{B_r}$. Hence $\sup_{B_r} |u| \leq \sup_{B_r} |\varphi| + \lambda r$.

By Lemma 10 $\sup_{B_r} |Du| \leq C$ with C only depending on r and $\sup_{B_R} |D\varphi|$, by the interior gradient estimates in [13], we can conclude that $u \in Lip(\overline{B_r})$ with $\|u\|_{Lip(\overline{B_r})}$ only depending on r, $\|\varphi\|_{L^\infty(\overline{B_R})}$, $\|D\varphi\|_{L^\infty(\overline{B_R})}$, and $\|DK\|_{L^\infty(\overline{B_R})}$. $\square$

5.4 Stability of Viscosity Solutions

It is a well known fact that the notion of viscosity solutions is stable with respect to uniform convergence, however we shall prove this stability property for our equation for reader convenience.

Lemma 12. *Let K_ε uniformly converge to K as ε goes to zero and let $\{u_\varepsilon\}$ be a sequence of viscosity solutions for the problem $\mathscr{L}(K_\varepsilon, u_\varepsilon) = 0$ which uniform converges to u as ε goes to zero, then u is a viscosity solution of the problem $\mathscr{L}(K, u) = 0$.*

Proof. We need to prove that u is either a viscosity sub-solution and super-solution. First we prove that u is a viscosity sub-solution. Let then $\xi_0 \in \Omega$ and let φ be a C^2 function in a neighborhood of ξ_0 such that φ touches u from above at ξ_0. We can choose a sequence $\xi_\varepsilon \to \xi_0$, as ε approaches zero, such that φ touches u_ε from above at ξ_ε. Since u_ε is a viscosity sub-solution for $\mathscr{L}(K_\varepsilon, u_\varepsilon) = 0$, then it holds $A(D\varphi, D^2\varphi)(\xi_\varepsilon) \geq 0$ and $\mathscr{L}(K_\varepsilon, u_\varepsilon)(\xi_\varepsilon) \geq 0$. Passing to the limit, as ε goes to zero, we get $A(D\varphi, D^2\varphi)(\xi_0) \geq 0$ and $\mathscr{L}(K, u)(\xi_0) \geq 0$. Therefore u is a viscosity sub-solution for $\mathscr{L}(K, u) = 0$.

Now, let $\xi_0 \in \Omega$ and let φ be a C^2 function in a neighborhood of ξ_0 such that φ touches u from below at ξ_0. We can choose a sequence $\xi_\varepsilon \to \xi_0$, as ε approaches zero, such that φ touches u_ε from below at ξ_ε.

In this situation, since u_ε is a viscosity super-solution for $\mathscr{L}(K_\varepsilon, u_\varepsilon) = 0$, we have to distinguish two cases.

- if for all $\varepsilon > 0$ $A(D\varphi, D^2\varphi)(\xi_\varepsilon) \geq 0$, then we have $\mathscr{L}(K_\varepsilon, u_\varepsilon)(\xi_\varepsilon) \leq 0$. Again passing to the limit as ε goes to zero, we get $A(D\varphi, D^2\varphi)(\xi_0) \geq 0$ and $\mathscr{L}(K_\varepsilon, u)(\xi_0) \leq 0$. Therefore u is a viscosity super-solution for $\mathscr{L}(K, u) = 0$.
- there exists a $\varepsilon > 0$ such that $A(D\varphi, D^2\varphi)(\xi_\varepsilon)$ is not positive semidefinite. Three situations can occur passing to the limit:

  (i) $A(D\varphi, D^2\varphi)(\xi_0) > 0$,
  (ii) $A(D\varphi, D^2\varphi)(\xi_0)$ is not positive semidefinite.
  (iii) $A(D\varphi, D^2\varphi)(\xi_0)$ has at least one null eigenvalue

  In (i) we can choose $\varepsilon > 0$ small enough such that $A(D\varphi, D^2\varphi)(\xi_\varepsilon) > 0$ and we have again the previous situation.
  In (ii) u is a viscosity super-solution for $\mathscr{L}(K, u) = 0$ at ξ_0 by definition.
  In (iii), since $\det A(D\varphi, D^2\varphi)(\xi_0) = 0$ and $K, F > 0$, we have $\mathscr{L}(K, \varphi)(\xi_0) = -K(\xi_0, \varphi)F(D\varphi) < 0$.

Therefore u is a viscosity super-solution for $\mathscr{L}(K, u) = 0$.   □

5.5   Proof of Theorem 16

The case $n = 1$ is contained in Theorem 6. Hence, we assume $n \geq 2$.
    We denote by $x = (x_1, x')$, $x' = (x_2, \ldots, x_n)$, $y = (y_1, y')$, $y' = (y_2, \ldots, y_n)$ with $x, y \in \mathbb{R}^n$, and $\xi = (x, y, t)$ is a point in $\mathbb{R}^{2n+1}$.
    We fix $R \in]0, 1[$ such that

$$\sup_{B_1 \times \mathbb{R}} K < \frac{1}{2R^n} \tag{42}$$

We are now going to provide a supersolution. For $0 \leq \sigma < 1$ and $0 < r < R$ we define

$$\phi_\sigma(\xi) = \phi_\sigma(x, y, t) := 2M(\sigma + |x'|^2 + |y'|^2)^\alpha,$$

where $\alpha = 1 - \frac{1}{n}$, and M a positive constant that will be determined later.
    Since $n \geq 2$, the exponent $\alpha = 1 - \frac{1}{n} \geq \frac{1}{2}$ and so that ϕ_σ is convex in $\mathbb{R}^{2n+1}$.
    Moreover, ϕ_σ is smooth for $\sigma > 0$, and independent of x_1, y_1 and t. From (36) we then obtain

$$A(D\phi_\sigma, D^2\phi_\sigma) = \Sigma \begin{pmatrix} D^2_{xx}\phi_\sigma & D^2_{xy}\phi_\sigma & 0 \\ D^2_{yx}\phi_\sigma & D^2_{yy}\phi_\sigma & 0 \\ 0 & 0 & 0 \end{pmatrix} \bar{\Sigma}^T$$

and since the first column and the first row of this matrix are null vectors we get $\det A(D\phi_\sigma, D^2\phi_\sigma) = 0$.

Therefore:

$$\mathscr{L}(K, \phi_\sigma) = -K(\,\cdot\,, \phi_\sigma)F(D\phi_\sigma) < 0 \quad \text{in } B_1, \quad \forall\, \sigma \in]0, 1[. \tag{43}$$

Thus, applying Lemma 40, the Dirichlet problem

$$\mathscr{L}(K, u) = 0 \quad \text{in } B_r, \qquad u = \phi_\sigma \quad \text{on } \partial B_r,$$

with $\sigma \in]0, 1[$ and $0 < r < R$, has a viscosity solution u_σ such that

$$\| u_\sigma \|_{L^\infty(\overline{B_r})} + \| u_\sigma \|_{Lip(\overline{B_r})} \leq C(r, \sigma, M)$$

with $C(r, \sigma, M)$ depending on σ only through $C(\phi_\sigma) := \|\phi_\sigma\|_{L^\infty(\overline{B_r})} + \|D\phi_\sigma\|_{L^\infty(\overline{B_r})}$.

On the other hand, an elementary computation shows that $C(\phi_\sigma) \leq 8M$. Then, we can choose $C(r, \sigma, M)$ independent of σ, and so

$$\| u_\sigma \|_{L^\infty(\overline{B_r})} + \| u_\sigma \|_{Lip(\overline{B_r})} \leq C(r, M). \tag{44}$$

We now define a supersolution.

Let

$$w_\sigma(x, y) = w_\sigma(x_1, x', y_1, y') := (r^2 + x_1^2 + y_1^2)(\sigma + |x'|^2 + |y'|^2)^\alpha, \tag{45}$$

where $\alpha = 1 - \frac{1}{n}$, and $\psi_\sigma(\xi) = \psi_\sigma(x, y, t) := M w_\sigma(x, y)$, with M a positive constant that will be determined in a moment.

Since $\phi_\sigma(\xi) = \phi_\sigma(x, y, t) := 2M(\sigma + |x'|^2 + |y'|^2)^\alpha$, we have $\psi_0 \leq \psi_\sigma \leq \phi_\sigma$, in B_1.

Claim. We claim that, if $0 < r < R$, we can fix $M = M(r)$ such that

$$\mathscr{L}(K, \psi_\sigma) > 0 \quad \text{in } B_r, \quad \forall\, \sigma \in]0, r^2[. \tag{46}$$

Assuming this claim for a moment, we can use the Comparison Principle to compare u_σ with ψ_σ and ϕ_σ. Indeed, by the claim, ϕ_σ and ψ_σ are, respectively, classical supersolution and subsolution to $\mathscr{L}(K, u) = 0$ in B_r. On the other hand $\psi_\sigma \leq \phi_\sigma$ in B_1, in particular, on ∂B_r. Thus, by the Comparison Principle,

$$\psi_\sigma \leq u_\sigma \leq \phi_\sigma \quad \text{in } B_r, \quad \forall \sigma \in]0, r^2[. \tag{47}$$

The uniform estimate (44) implies the existence of a sequence $\sigma_j \searrow 0$ such that $(u_{\sigma_j})_{j \in \mathbb{N}}$ uniformly converges to a viscosity solution $u \in Lip(\overline{B_r})$ to the Dirichlet problem

$$\mathcal{L}(K, u) = 0 \quad \text{in } B_r, \quad u = \phi_0 \text{ on } \partial B_r.$$

Moreover, from (47), we get

$$\psi_0 \leq u \leq \phi_0 \quad \text{in } B_r.$$

In particular

$$M r^2 |x_2|^{2\alpha} \leq u(0, x_2, 0, \ldots, 0) \leq 2M |x_2|^{2\alpha}. \tag{48}$$

These inequalities imply:

- $u \notin C^1$, if $2\alpha = 1$ (i.e. $n = 2$)
- $u \notin C^{1,\beta}$, for every $\beta > 2\alpha - 1 = 1 - \frac{2}{n}$ if $2\alpha > 1$ (i.e. $n > 2$).

The first statement is trivial.

For the second one we only have to remark that if $2\alpha > 1$, then $\partial_{x_2} u(0, 0, \ldots, 0) = 0 = u(0, 0, \ldots, 0)$ so that, if u would be $C^{1,\beta}$, with $\beta > 2\alpha - 1$, we would have $u(0, x_2, \ldots, 0) \leq C |x_2|^{1+\beta}$ for a suitable $C > 0$ and for every x_2 sufficiently small. Hence, by the first inequality in (48), it would be $\beta \leq 2\alpha - 1$, a contradiction.

Proof (Proof of the claim). Elementary direct computations show that

$$|Dw_\sigma|^2 = 4\Big((x_1^2 + y_1^2)(\sigma + |x'|^2 + |y'|^2)^{2\alpha}$$
$$+ \alpha^2 (|x'|^2 + |y'|^2)(r^2 + x_1^2 + y_1^2)^2 (\sigma + |x'|^2 + |y'|^2)^{2(\alpha-1)} \Big)$$

and

$$\det A(Dw_\sigma, D^2 w_\sigma) = 2^{2n} f_\sigma \tag{49}$$

with

$$f_\sigma = \alpha^n (r^2 + x_1^2 + y_1^2)^{n-2} \frac{r^2(\alpha^{-1}\sigma + |x'|^2 + |y'|^2) + \alpha^{-1}\sigma(|x'|^2 + |y'|^2)}{(\sigma + |x'|^2 + |y'|^2)}.$$

Then, for every $\sigma \in]0, r^2[$,

$$|Dw_\sigma|^2 \leq 2^{2\alpha+3} r^{4\alpha+2} \quad \text{in } B_r, \tag{50}$$

and

$$f_\sigma \geq \alpha^n \, r^{2(n-1)} \quad \text{in } B_r. \tag{51}$$

Keeping in mind that $\psi_\sigma = M w_\sigma$ is independent of t, we get

$$\frac{\det A(D\psi_\sigma, D^2\psi_\sigma)}{F(D\psi_\sigma)} = \frac{(2M)^n f_\sigma}{(1 + M^2|Dw_\sigma|^2)^{\frac{n+2}{2}}}.$$

Therefore from (51) and (50), we obtain

$$\frac{\det A(D\psi_\sigma, D^2\psi_\sigma)}{F(D\psi_\sigma)} \geq \frac{(2M\alpha)^n \, r^{2(n-1)}}{(1 + 2^{2\alpha+3} r^{4\alpha+2} M^2)^{\frac{n+2}{2}}} \quad \text{in } B_r. \tag{52}$$

Choosing $M = 2^{-\alpha-(3/2)} r^{-2\alpha-1}$, the right hand side of (52) equals

$$\frac{(2\alpha)^n \, 2^{-(3/2)n+1} r^{-n}}{2^{\frac{n+2}{2}}} = \frac{C(n)}{r^n} > \frac{1}{2R^n}, \quad \text{if } r < (2C(n))^{1/n} R.$$

Then from (42) we obtain

$$\mathscr{L}(K, \psi_\sigma) = \det A(D\psi_\sigma, D^2\psi_\sigma) - K(\cdot, \psi_\sigma) F(D\psi_\sigma)$$

$$> F(D\psi_\sigma)\left(\frac{1}{2R^n} - K(\cdot, \psi_\sigma)\right) > 0.$$

This proves the claim (46) and completes the proof of the theorem.   □

6 Symmetry Results

The study of surfaces in the Euclidean space with either constant Gauss curvature or constant mean curvature received in the past a great amount of attention. In 1899 Liebmann [26] proved that the spheres are the only compact surfaces in $\mathbb{R}^3$ with constant Gauss curvature. In 1952 Süss [41] extended the Liebmann result showing that a compact convex hypersurface in the Euclidean space must be a sphere, provided that for some j the j-th elementary symmetric polynomial in the principal curvatures is constant. In 1954 Hsiung [22] proved that the "convexity" assumption can be relaxed to the "star-shapedness" one. The proofs of the above results are based on the Minkowski formulae. A breakthrough for this sort of problems was made by Alexandrov [1] in 1956, who proved that the sphere is the only compact hypersurface embedded into the Euclidean space with constant mean curvature. Alexandrov method is completely different from the Liebmann-Süss method, and is based on the moving plane technique, on the interior maximum principle for elliptic equations and on the boundary maximum principle of Hopf type for uniformly elliptic equations. In 1978 Reilly [37] obtained another proof of the Alexandrov theorem combining the Minkowski formulae with some new elegant arguments.

In a joint paper with Lanconelli [31] we proved a strong comparison principle, leading to symmetry theorems for domains with constant curvatures.

Precisely:

Theorem 17. *Let $D \subseteq \mathbb{C}^{n+1}$ be a strictly j-pseudoconvex domain with connected boundary, $1 \leq j \leq n$. Let $B_R(z_0) \subseteq D$ be a tangent sphere to ∂D at some point $p \in \partial D$. If $K_{\partial D}^{(j)}(z)$ is the j-th Levi curvature of ∂D at $z \in \partial D$ and*

$$K_{\partial D}^{(j)}(z) \geq 1/R^j, \quad \forall z \in \partial D,$$

then $D = B_R(z_0)$.

In [31] we also proved that if Ω is a bounded domain of $\mathbb{C}^{n+1}$, with boundary a real hypersurface of class C^2, then the j-th Levi curvature of $\partial \Omega$ at $z = (z_1, \ldots, z_{n+1}) \in \partial \Omega$ writes in term of defining function f of $\Omega = \{f(z) < 0\}$ as

$$K_{\partial \Omega}^{(j)}(z) = -\frac{1}{\binom{n}{j}} \frac{1}{|\partial f|^{j+2}} \sum_{1 \leq i_1 < \cdots < i_{j+1} \leq n+1} \Delta_{(i_1, \cdots, i_{j+1})}(f) \tag{53}$$

for all $j = 1, \ldots, n$, where $|\partial f| = \sqrt{\sum_{j=1}^{n+1} |f_j|^2}$

$$\Delta_{(i_1, \cdots, i_{j+1})}(f) = \det \begin{pmatrix} 0 & f_{\bar{i}_1} & \cdots & f_{\bar{i}_{j+1}} \\ f_{i_1} & f_{i_1, \bar{i}_1} & \cdots & f_{i_1, \bar{i}_{j+1}} \\ \vdots & \vdots & \ddots & \vdots \\ f_{i_{j+1}} & f_{i_{j+1}, \bar{i}_1} & \cdots & f_{i_{j+1}, \bar{i}_{j+1}} \end{pmatrix} \tag{54}$$

and $f_j = \frac{\partial f}{\partial z_j}$, $f_{\bar{j}} = \overline{f_j}$, $f_{j\bar{\ell}} = \frac{\partial^2 f}{\partial z_j \partial \bar{z}_\ell}$.

Theorem 17 suggested the following question: are spheres the unique compact hypersurfaces with constant Levi curvatures? Klingenberg in [23] gave a first positive answer to this problem by showing that a compact and strictly pseudoconvex real hypersurface $M \subset \mathbb{C}^{n+1}$ is isometric to a sphere, provided that M has constant horizontal mean curvature, the Levi form is diagonal and the characteristic direction of the hypersurface is a geodesic.

Later on in [29] we relaxed Klingerberg conditions and we proved that if the characteristic direction is a geodesic, then Alexandrov Theorem holds for hypersurfaces with positive constant Levi mean curvature (see Theorem 24).

The problem of characterizing hypersurfaces with constant Levi curvature has been studied by many authors. Hounie and Lanconelli in [20] showed the result for Reinhardt domain of $\mathbb{C}^2$, i.e. for domains D such that if $(z_1, z_2) \in D$ then

$(e^{i\theta_1}z_1, e^{i\theta_2}z_2) \in D$ for all real θ_1, θ_2. Under this hypothesis, in a neighborhood of a point, there is a defining function F only depending on the radius $r_1 = |z_1|$, $r_2 = |z_2|$, $F(r_1, r_2) = f(r_2^2) - r_1^2$ with f the solution of the ODE

$$sff'' = sf'^2 - k(f + sf'^2)^{3/2} - ff' \tag{55}$$

Alexandrov Theorem follows from uniqueness of the solution of (55). Their technique has then been used in [21] to prove an Alexandrov Theorem for Reinhardt domains in $\mathbb{C}^{n+1}$ with spherical symmetry for every n. Then in [33] Monti and Morbidelli proved a Darboux -type theorem for $n \geq 2$: the unique Levi umbelical hypersurfaces in $\mathbb{C}^{n+1}$ with all constant Levi curvatures are spheres or cylinders.

In [28] we proved some integral formulas for compact hypersurfaces, of independent interest, and then we follow the Reilly approach to prove the isoperimetric estimate Theorem 22 and the following Alexandrov type result.

Theorem 18 (An **Alexandrov type Theorem**). *Let $\Omega \subset \mathbb{C}^2$ be a bounded star-shaped domain with boundary a smooth real hypersurface. If the j-Levi curvature is a positive constant K at every point of $\partial\Omega$ and the maximum of the mean curvature of $\partial\Omega$ is bounded from above by K, then $\partial\Omega$ is a sphere and Ω is a ball.*

In this section we shall use the null Lagrangian property for elementary symmetric functions in the eigenvalues of the complex Hessian matrix and the classical divergence theorem to prove an integral formula (59) for a closed hypersurface in term of the j-th Levi curvature. We prove an estimate of the j-th Levi curvature (61) and we use the Minkowski formula for the classical mean curvature to conclude the proof of Theorem 18.

Then, in the last part of this section, we shall introduce differential geometry techniques to show a second result of Alexandrov type, Theorem 24.

6.1 Examples and Open Problems

Let us begin with some examples

Example 6 (m-Levi curvature of a sphere). $f(z) = |z|^2 - r^2$ is a defining function for B_r, and the Levi form of ∂B_r is

$$\mathscr{L}_p(\partial B_r) = (1/r)I_n, \quad \forall p \in \partial B_r,$$

The eigenvalues of the Levi form are all equal to $1/r$, thus

$$K_{\partial B_r}^{(m)}(p) = 1/r^m, \quad \forall p \in \partial B_r.$$

Problem 1. Let $D \subset \mathbb{C}^{n+1}$ be a smooth, bounded and connected domain. Assume

$$K_{\partial D}^{(n)}(p) = c \ \text{(constant)} \quad \text{for every } p \in \partial D$$

Then: is it true that D is a ball?

   That is: does there exist $\alpha \in \mathbb{C}^{n+1}$ and $R > 0$ such that

$$D = B_R(\alpha) \ ?$$

If yes : $c = \left(\frac{1}{R}\right)^n > 0$!

Remark 12. The boundedness assumption is necessary.

Example 7 (Cylinders with constant and positive Gauss-Levi curvature). Let us now consider the cylinder

$$C_R = \{(z_1, \cdots, z_{n+1}) \in \mathbb{C}^{n+1} : (Re \, z_1)^2 + \cdots + (Re \, z_{n+1})^2 < R^2\}.$$

A defining function of C_R is

$$f(z) = (Re \, z_1)^2 + \cdots + (Re \, z_{n+1})^2 - R^2.$$

The Gauss-Levi curvature is:

$$K_{\partial C_R}(p) = \frac{1}{(2R)^n}, \quad \forall p \in \partial C_R.$$

6.2 Ingredients of Alexandrov Method

Alexandrov proved that every direction is a direction of symmetry by using the following ingredients:

1. The prescribed mean curvature is invariant with respect to hyperplane reflections.
2. The equation is uniformly elliptic and the strong comparison principle holds.
3. Boundary comparison principle.

6.3 An Obstacle to the Boundary Comparison Principle

In general sums of squares of Hörmander vector fields do not satisfy Hopf Lemma, as the following example shows

Example 8 (Hounie, Lanconelli). In $\mathbb{R}^2$ consider $L = X^2 + Y^2$

$$X = y\frac{\partial}{\partial x}, \quad Y = \frac{\partial}{\partial y} - y\frac{\partial}{\partial x} \Rightarrow [X, Y] = -\frac{\partial}{\partial x}.$$

The function $u(x, y) = y^4 - 6(x + y^2/2)^2$ satisfies

1. $Lu = 0$
2. $u < 0$ in the half plane $x \geq 0$ except the origin and $u(0, 0) = 0$. Then the restriction of u to the half plane $x \geq 0$ has a strict maximum at the origin
3. $\nabla u(0, 0) = 0$.

Hence, L does not satisfy Hopf Lemma.

6.4   Strong Comparison Principle

Even if the prescribed Levi curvature equations are not elliptic at any point, in [31] we proved the following strong comparison principle.

Theorem 19 (STRONG COMPARISON PRINCIPLE). *Let D and D' be strictly m-pseudoconvex domains of $\mathbb{C}^{n+1}$ with connected boundaries. If*

1. *$D' \subseteq D$ and $\partial D \cap \partial D' \neq \varnothing$*
2. *$K_{\partial D'}^{(m)}(p') \leq K_{\partial D}^{(m)}(p)$, $\forall \, p \in \partial D$ and $p' \in \partial D'$*

then $D' = D$.

Proof. We just give a sketch of the proof and we suggest the interested reader to see the original proof in [31] for details. We locally write ∂D and $\partial D'$ as the graph of u and $v \in C^2$, respectively. Then, the function $w = u - v \leq 0$ in an open set Ω and with the notation of Sect. 4.7.1

$$\mathscr{L}^m(w) = \sum_{i,j=1}^{2n} a_{ij}(x)Z_i Z_j w + b_j(x)Z_j w \geq 0,$$

where

1. the matrix $a_{ij}(x) = a_{ji}(x)$ and it is positive definite as a quadratic form at every point $x \in \Omega$
2. there is a couple (i, j) such that $[Z_i, Z_j] \neq 0$.   □

We then conclude by the following Maximum principle of Bony type

Theorem 20 (MAXIMUM PRINCIPLE). *Let $\Omega_0 \subseteq \Omega$ be an open and connected set. Let $w \in C^2(\Omega_0, \mathbb{R})$ such that*

$$\begin{cases} \mathscr{L}^m w \geq 0 \text{ in } \Omega_0 \\ w \leq 0 \quad \text{ in } \Omega_0 \end{cases}$$

Then $w < 0$ in Ω_0 or $w \equiv 0$ in Ω_0.

By the strong comparison principle we get the proof of the identification result Theorem 17.

6.5 Null Lagrangian Property for Elementary Symmetric Functions in the Eigenvalues of the Complex Hessian Matrix

Given a Hermitian matrix A, let $\sigma_j(A)$ be the j-th elementary symmetric function in the eigenvalues of A.

Let us recall that if A is the $(n+1) \times (n+1)$ matrix with eigenvalues $\lambda_1, \ldots, \lambda_{n+1}$ then

$$\sigma_j(A) = \sum_{1 \leq i_1 < \cdots < i_j \leq n+1} \lambda_{i_1} \cdots \lambda_{i_j}.$$

If we choose $A = [a_{\ell\bar{k}}] = \partial\bar{\partial} f$ the complex Hessian matrix of a smooth function f and we denote by $\frac{\partial\sigma_j(A)}{\partial a_{\ell\bar{k}}}$ the partial derivative of the function σ_j with respect to the term of place $\ell\bar{k}$ of the matrix A, then the following *null Lagrangian property* holds

$$\sum_\ell \partial_\ell \left(\frac{\partial\sigma_{j+1}(\partial\bar{\partial} f)}{\partial a_{\ell\bar{k}}} \right) = 0, \forall k = 1, \ldots, n+1. \tag{56}$$

Moreover, by elementary linear algebra, for every $f \in C^2$ and for every $j = 1, \ldots, n+1$

$$\sum_{\ell,k=1}^{n+1} \frac{\partial\sigma_{j+1}}{\partial a_{\ell\bar{k}}}(\partial\bar{\partial} f) f_\ell f_{\bar{k}} = -\sum_{1 \leq i_1 < \cdots < i_{j+1} \leq n+1} \Delta_{(i_1,\ldots,i_{j+1})}(f) \tag{57}$$

where

$$\Delta_{(i_1,\cdots,i_{j+1})}(f) = \det \begin{pmatrix} 0 & f_{\bar{i}_1} & \cdots & f_{\bar{i}_{j+1}} \\ f_{i_1} & f_{i_1,\bar{i}_1} & \cdots & f_{i_1,\bar{i}_{j+1}} \\ \vdots & \vdots & \ddots & \vdots \\ f_{i_{j+1}} & f_{i_{j+1},\bar{i}_1} & \cdots & f_{i_{j+1},\bar{i}_{j+1}} \end{pmatrix} \tag{58}$$

and $f_j = \frac{\partial f}{\partial z_j}$, $f_{\bar{j}} = \overline{f_j}$, $f_{j\bar{\ell}} = \frac{\partial^2 f}{\partial z_j \partial \bar{z}_\ell}$.

6.6 Integral Formulas

By using the null Lagrangian property for elementary symmetric functions in the eigenvalues of the complex Hessian matrix and the classical divergence Theorem we get our integral formulas for closed hypersurfaces.

Theorem 21. *Let Ω be a bounded domain of $\mathbb{C}^{n+1}$ with boundary a real hypersurface of class C^2. For every defining function f of class C^2 of $\Omega = \{f(z) < 0\}$ and for every $j = 1, \ldots, n$ by the divergence theorem we have*

$$\int_{\Omega} \sigma_{j+1}(\partial\bar{\partial}f)dx = \binom{n+1}{j+1}\frac{1}{2(n+1)}\int_{\partial\Omega} K_{\partial\Omega}^{(j)}(z)|\partial f|^{j+1}d\sigma(x), \qquad (59)$$

where $K_{\partial\Omega}^{(j)}$ is the j-th Levi curvature of $\partial\Omega$.

Proof. Since σ_j is a homogenous function of degree j, i.e. $\sigma_j(tA) = t^j\sigma_j(A)$ for every real t, we get

$$\sigma_{j+1}(\partial\bar{\partial}f) = \frac{1}{j+1}\sum_{\ell,k=1}^{n+1}\frac{\partial\sigma_{j+1}(\partial\bar{\partial}f)}{\partial a_{\ell\bar{k}}}f_{\ell\bar{k}} \qquad (60)$$

Let us set $v_\ell = \frac{\partial_\ell f}{|\partial f|}$ and let us identify $z \in \mathbb{C}^{n+1}$ with $x \in \mathbb{R}^{2(n+1)}$, then by the divergence Theorem plus the null Lagrangian property (56), together with (57), (58) and (54), (53), we get

$$\int_{\Omega} \sigma_{j+1}(\partial\bar{\partial}f)dx = \frac{1}{j+1}\int_{\Omega}\sum_{\ell,k=1}^{n+1}\partial_\ell\left(\frac{\partial\sigma_{j+1}}{\partial a_{\ell\bar{k}}}(\partial\bar{\partial}f)f_{\bar{k}}\right)dx$$

$$= \frac{1}{2(j+1)}\int_{\partial\Omega}\sum_{\ell,k=1}^{n+1}\left(\frac{\partial\sigma_{j+1}}{\partial a_{\ell\bar{k}}}(\partial\bar{\partial}f)f_{\bar{k}}v_\ell\right)d\sigma(x)$$

$$= \frac{1}{2(j+1)}\int_{\partial\Omega}\sum_{\ell,k=1}^{n+1}\frac{\left(\frac{\partial\sigma_{j+1}}{\partial a_{\ell\bar{k}}}(\partial\bar{\partial}f)f_{\bar{k}}f_\ell\right)}{|\partial f|}d\sigma(x)$$

$$= -\frac{1}{2(j+1)}\int_{\partial\Omega}\frac{\sum_{1\leq i_1<\cdots<i_{j+1}\leq n+1}\Delta_{(i_1,\cdots,i_{j+1})}(f)}{|\partial f|}d\sigma(x)$$

$$= \binom{n}{j}\frac{1}{2(j+1)}\int_{\partial\Omega}K_{\partial\Omega}^{(j)}(z)|\partial f|^{j+1}d\sigma(x). \qquad \square$$

6.7   Isoperimetric Inequality

Theorem 22 (ISOPERIMETRIC ESTIMATE). *Let Ω be a bounded domain of $\mathbb{C}^{n+1}$ with boundary a real hypersurface of class C^∞. If $K_{\partial\Omega}^{(j)}$ is non negative at every point of $\partial\Omega$ then*

$$\int_{\partial\Omega} \left(\frac{1}{K_{\partial\Omega}^{(j)}(x)} \right)^{1/j} d\sigma(x) \geq 2(n+1)|\Omega| \tag{61}$$

where $|\Omega|$ is the Lebesgue measure of Ω.

Moreover, if $K_{\partial\Omega}^{(j)}$ is constant, then the equality holds in (61) if and only if Ω is a ball of radius $\left(\frac{1}{K_{\partial\Omega}^{(j)}} \right)^{1/j}$.

Let us remark that there are non spherical sets which satisfy the equality in (61) (see (62)).

Example 9. In $\mathbb{C}^2$ let us take the function

$$f(z) = -R^2 + |z|^2 + \frac{\mathrm{Re}\,(z_1^2 + z_2^2)}{2} \tag{62}$$

as a defining function for Ω. Then the set of zeros of the function f is not a sphere, it is an ellipsoid. However, equality holds in our isoperimetric inequality because the function $\mathrm{Re}\,(z_1^2 + z_2^2)$ is a pluriharmonic function and therefore $\partial\bar\partial f = Id$.

Thus, the class of sets which satisfy the equality in (61) is larger than the class of sets which satisfy the equality in the classical isoperimetric inequality and in the Alexandrov Fenchel inequalities for quermassintegrals (see [16] and [43]).

Proof. We just give a sketch of the proof. The ingredients are the following

1. For every defining function $f \in C^2(\bar\Omega)$

$$\int_\Omega \sigma_{j+1}(\partial\bar\partial f)dx = \binom{n+1}{j+1} \frac{1}{2(n+1)} \int_{\partial\Omega} K_{\partial\Omega}^{(j)}(z)|\partial f|^{j+1}d\sigma(x).$$

2. (HÖLDER INEQUALITY) For every defining function $f \in C^2(\bar\Omega)$

$$\int_{\partial\Omega} K^{(j)}|\partial f|^{j+1}d\sigma(x) \geq \frac{\left(\int_{\partial\Omega} |\partial f|d\sigma(x) \right)^{j+1}}{\left(\int_{\partial\Omega} \left(\frac{1}{K^{(j)}} \right)^{1/j} d\sigma(x) \right)^j}$$

$$= \frac{(2 \int_\Omega \mathrm{trace}\,(\partial\bar\partial f)dx)^{j+1}}{\left(\int_{\partial\Omega} \left(\frac{1}{K^{(j)}} \right)^{1/j} d\sigma(x) \right)^j}.$$

3. (NEWTON INEQUALITY)

$$\sigma_j(\partial\bar\partial f) \le \binom{n+1}{j}\left(\frac{\text{trace } \partial\bar\partial f}{n+1}\right)^j$$

for all $j = 2,\ldots,n+1$, with equality holding iff $(\partial\bar\partial f)$ is proportional to Id.

4. Let us choose $f : \bar\Omega \to \mathbb{R}$ be the solution of the Dirichlet problem

$$\begin{cases} \text{trace } \partial\bar\partial f = 1, \text{ in } \Omega; \\ f = 0, \qquad\quad \text{on } \partial\Omega. \end{cases}$$

trace $\partial\bar\partial = \frac{1}{4}\Delta$, with Δ the Laplacian operator in $\mathbb{R}^{2(n+1)}$.

Then, by Hopf Lemma $|\partial f| \ne 0$ on $\partial\Omega$ and f is a defining function for the domain Ω. With this particular choice of f we have

$$\int_\Omega \sigma_{j+1}(\partial\bar\partial f)dx = \binom{n+1}{j+1}\frac{1}{2(n+1)}\int_{\partial\Omega} K_{\partial\Omega}^{(j)}(z)|\partial f|^{j+1}d\sigma(x)$$

$$\ge \binom{n+1}{j+1}\frac{1}{2(n+1)}\frac{(2|\Omega|)^{j+1}}{\left(\int_{\partial\Omega}\left(\frac{1}{K^{(j)}}\right)^{1/j}d\sigma(x)\right)^j}$$

and

$$\int_\Omega \sigma_{j+1}(\partial\bar\partial f)dx \le \binom{n+1}{j+1}\frac{1}{(n+1)^{j+1}}\int_\Omega (\text{trace } (\partial\bar\partial f))^{j+1}\,dx$$

$$= \binom{n+1}{j+1}\frac{|\Omega|}{(n+1)^{j+1}}.$$

By putting together the two above estimates we finally get

$$\frac{(2|\Omega|)^j}{\left(\int_{\partial\Omega}\left(\frac{1}{K^{(j)}}\right)^{1/j}d\sigma(x)\right)^j} \le \frac{1}{(n+1)^j}$$

and

$$\int_{\partial\Omega}\left(\frac{1}{K^{(j)}}\right)^{1/j}d\sigma(x) \ge 2(n+1)|\Omega|. \qquad\qquad \square$$

In particular, if equality holds in the isoperimetric estimate and $K^{(j)}$ is a positive constant then $\partial\bar{\partial}f = \frac{1}{n+1}Id$ in $\bar{\Omega}$ and $|\partial f| = \frac{1}{(n+1)(K^{(j)})^{1/j}}$ is constant on $\partial\Omega$.

Then the Dirichlet problem is over-determinate and by Serrin's Theorem [38] we can conclude that Ω is a ball and $\partial\Omega$ is a sphere.

6.8   First Alexandrov Type Result

If Ω is a starlike domain then by the MINKOWSKI FORMULA

$$\text{meas}\,(\partial\Omega) = \int_{\partial\Omega} d\sigma(x) = \int_{\partial\Omega} H(x)\langle x, \nu(x)\rangle d\sigma(x)$$
$$\leq 2(n+1)|\Omega|\max_{\partial\Omega} H$$

where H is the mean curvature of $\partial\Omega$ and ν is the outward unit normal. Then

$$\frac{\text{meas}\,(\partial\Omega)}{2(n+1)|\Omega|} \leq \max_{\partial\Omega} H$$

On the other side, if $K^{(j)}_{\partial\Omega}$ is constant, then

$$\left(K^{(j)}_{\partial\Omega}\right)^{1/j} \leq \frac{\text{meas}\,(\partial\Omega)}{2(n+1)|\Omega|} \tag{63}$$

and an Alexandrov type theorem holds for star-shaped domain whose classical mean curvature is bounded from above by a constant j-Levi curvature to the power $1/j$. This concludes the proof of Theorem 18.

6.9   Curvature Lines and Geodesics. Second Alexandrov Type Result

By using the notation of Sect. 2, let S be the Weingarten operator (or shape operator), defined by

$$S(V) = -\nabla_V \mathcal{N}, \ \forall \ V \in TM \tag{64}$$

Definition 25. Let $V \in TM$. V is a eigenvector for S if there exists a function (eigenvalue) $\lambda : M \to \mathbb{R}$ such that $S(V) = \lambda V$ on M.

Let γ be the integral curve of V, i.e. $\gamma \subseteq M$ is a line such that $\dot{\gamma} = V$. If V is a eigenvector for S then we refer to γ as a curvature line.

If V is unitary, then the value of λ is $\lambda = \mathbb{II}(V, V)$ because

$$\mathbb{II}(V, V) = \langle S(V), V \rangle = \langle \lambda V, V \rangle = \lambda \langle V, V \rangle = \lambda$$

Definition 26. Let $V \in TM$. The integral curve of V is a geodesic if $\nabla_V V \in \mathbb{R}\mathcal{N}$, i.e. if the field $\nabla_V V$ is normal to M.

This definition of geodesic coincides with that one of minimizing curve for the distance functional $d_{p,q}(\gamma)$, induced by the metric of $\mathbb{C}^{n+1}$, i.e. if $p, q \in M$, for all curves $\gamma : [t_1, t_2] \to M$ such that $\gamma(t_1) = p$ and $\gamma(t_2) = q$

$$d_{p,q}(\gamma) = \int_{t_1}^{t_2} \sqrt{\langle \dot{\gamma}, \dot{\gamma} \rangle} dt$$

and the geodesic is the curve that realizes $min(d_{p,q}(\gamma))$.

With an abuse of language, we will also refer to the vector field V as a curvature line or a geodesic if the corresponding integral curve is a curvature line or a geodesic respectively.

Lemma 13. *Let T be the characteristic direction of M. T is a curvature line if and only if it is a geodesic.*

Proof. If T is a curvature line, one has

$$S(T) = \lambda T, \quad \lambda = \mathbb{II}(T, T) \tag{65}$$

For all $X \in HM$,

$$\langle \nabla_T \mathcal{N}, X \rangle = \langle -S(T), X \rangle = -\mathbb{II}(T, T)\langle T, X \rangle = 0 \tag{66}$$

For all $X \in HM$ we have

$$0 = \langle \nabla_T \mathcal{N}, X \rangle = \langle J(\nabla_T \mathcal{N}), J(X) \rangle = \langle \nabla_T T, J(X) \rangle \tag{67}$$

Moreover T is unitary and by differentiating along T one has

$$\langle \nabla_T T, T \rangle = 0 \tag{68}$$

Therefore, by using (67) and (68) it is proved that $\nabla_T T \in \mathbb{R}\mathcal{N}$.

To prove the converse invert the above procedure. $\square$

By using Codazzi equations and Chow Theorem, in [29] we showed a characterization result for non Levi flat real smooth hypersurfaces in $\mathbb{C}^{n+1}$, whose unit characteristic direction T is a geodesic.

Theorem 23 (Martino, M. 2010). *Let M be a non Levi flat real hypersurface in $\mathbb{C}^{n+1}$. If the characteristic direction T is a geodesic for M, then $\mathbb{III}(T, T)$ is constant.*

The above Theorem cannot be inverted. Indeed, in [29] we showed a non Levi flat hypersurface whose characteristic direction is not a geodesic, but $\mathbb{III}(T, T)$ is constant.

As an application of the previous Theorem we get a result of characterization of spheres, of Alexandrov type:

Theorem 24. *Let M be a compact real hypersurface in $\mathbb{C}^{n+1}$ with positive constant Levi mean curvature. If the characteristic direction T is a geodesic for M, then M is a sphere.*

Proof. If M has constant positive Levi mean curvature, then M is non Levi flat, and since T is a curvature line, one has that $\mathbb{III}(T, T)$ is constant on M. Then the mean curvature of M is constant. By using the compactness of M and by the classical Alexandrov's theorem we get that M is a sphere. □

References

1. A.D. Alexandrov, Uniqueness theorems for surfaces in the large I. Vestnik Leningrad Univ. **11**, 5–17 (1956)
2. E. Bedford, B. Gaveau, Hypersurfaces with bounded Levi form. Indiana Univ. J. **27**(5), 867–873 (1978)
3. E. Bedford, B. Gaveau, Envelopes of Holomorphy of certain 2-spheres in $\mathbb{C}^2$. Am. J. Math. **105**(4), 975–1009 (1983)
4. E. Bedford, W. Klingenberg, On the envelope of holomorphy of a 2-sphere in $\mathbb{C}^2$. J. Am. Math. Soc. **4**(3), 623–646 (1991)
5. A. Boggess, *CR Manifolds and the Tangential Cauchy-Riemann Complex*. Studies in Advanced Mathematics (CRC Press, Boca Raton, 1991)
6. G. Citti, A comparison theorem for the Levi equation. Rend. Mat. Accad. Lincei **4**, 207–212 (1993)
7. G. Citti, C^∞ regularity of solutions of the Levi equation. Ann. Inst. H. Poincaré Anal. Non Linéaire **15**, 517–534 (1998)
8. G. Citti, E. Lanconelli, A. Montanari, Smoothness of Lipschitz continuous graphs with non vanishing Levi curvature. Acta Math. **188**(1), 87–128 (2002)
9. G. Citti, A. Montanari, Strong solutions for the Levi curvature equation. Adv. Differ. Equat. **5**(1–3), 323–342 (2000)
10. G. Citti, A. Montanari, Analytic estimates for solutions of the Levi equation. J. Differ. Equat. **173**(2), 356–389 (2001)
11. G. Citti, A. Montanari, Regularity properties of solutions of a class of elliptic-parabolic nonlinear Levi type equations. Trans. Am. Math. Soc. **354**(7), 2819–2848 (2002)
12. M.G. Crandall, H. Ishii, P.L. Lions, User's guide to viscosity solutions of second order partial differential equations. Bull. Am. Math. Soc. **27**, 1–67 (1992)
13. F. Da Lio, A. Montanari, Existence and uniqueness of Lipschitz continuous graphs with prescribed Levi curvature. Ann. Inst. H. Poincaré Anal. Non Linéaire **23**(1), 1–28 (2006)
14. A. Debiard, B. Gaveau, Problème de Dirichlet pour l'équation de Levi. Bull. Sci. Math. (2) **102**(4), 369–386 (1978)

15. S. Dragomir, G. Tomassini, *Differential Geometry and Analysis on CR Manifolds*. Progress in Mathematics (Birkhaüser, Boston, 2006)
16. H. Federer, *Geometric Measure Theory*. Die Grundlehren der mathematischen Wissenschaften, Band 153 (Springer, New York, 1969)
17. C.E. Gutiérrez, *The Monge-Ampère Equation*. Progress in Nonlinear Differential Equations and Their Applications (Birkhaüser, Boston, 2001)
18. C.E. Gutiérrez, E. Lanconelli, A. Montanari, Nonsmooth hypersurfaces with smooth Levi curvature. Nonlinear Anal. TMA **76**, 115–121 (2013)
19. L. Hörmander, *Notions of Convexity*. Progress in Mathematics, vol. 127 (Birkhäuser, Boston, 1994)
20. J. Hounie, E. Lanconelli, An Alexandrov type Theorem for Reinhardt domains of $\mathbb{C}^2$. Contemp. Math. **400**, 129–146 (2006)
21. J. Hounie, E. Lanconelli, A sphere theorem for a class of Reinhardt domains with constant Levi curvature. Forum Math. **20**, 571–586 (2008)
22. C.C. Hsiung, Some integral formulas for closed hypersurfaces. Math. Scand. **2**, 286–294 (1954)
23. W. Klingenberg, Real hypersurfaces in Käler manifolds. Asian J. Math. **5**(1), 1–17 (2001)
24. S. Krantz, *Function Theory of Several Complex Variables* (Wiley, New York, 1982)
25. E. Lanconelli, A. Montanari, *On a Class of Fully Nonlinear PDEs from Complex Geometry*. Contemporary Mathematics, vol. 594, pp. 231–242. AMS volume dedicated to Professor Patrizia Pucci on the occasion of her 60th birthday (2013), http://dx.doi.org/10.1090/conm/594/11796
26. H. Liebmann, *Eine neue eigenschaft der Kugel*. (German) Gött. Nachr. 44–55 (1899)
27. V. Martino, A. Montanari, Local Lipschitz continuity of graphs with prescribed Levi mean curvature. NoDEA Nonlinear Differ. Equat. Appl. **14**(3–4), 377–390 (2007)
28. V. Martino, A. Montanari, Integral formulas for a class of curvature PDE's and applications to isoperimetric inequalities and to symmetry problems. Forum Math. **22**, 253–265 (2010)
29. V. Martino, A. Montanari, On the characteristic direction of real hypersurfaces in $\mathbb{C}^{n+1}$, and a symmetry result. Adv. Geom. **10**(3), 371–377 (2010)
30. A. Montanari, On the regularity of solutions of the prescribed Levi curvature equation in several complex variables. Nonlinear elliptic and parabolic equations and systems (Pisa, 2002). Comm. Appl. Nonlinear Anal. **10**(2), 63–71 (2003)
31. A. Montanari, E. Lanconelli, Pseudoconvex fully nonlinear partial differential operators: strong comparison theorems. J. Differ. Equat. **202**, 306–331 (2004)
32. A. Montanari, F. Lascialfari, The Levi Monge-Ampère equation: smooth regularity of strictly Levi convex solutions. J. Geom. Anal. **14**(2), 331–353 (2004)
33. R. Monti, D. Morbidelli, Levi umbilical surfaces in complex space. J. Reine Angew. Math. **603**, 113–131 (2007)
34. A.V. Pogorelov, The Dirichlet problem for the n-dimensional analogue of the Monge-Ampère equation. Dokl. Akad. Nauk SSSR **201**, 790–793 (1971) (Russian). English translation in Soviet Math. Dokl. **12**, 1727–1731 (1971)
35. R.M. Range, *Holomorphic Functions and Integral Representation Formulas in Several Complex Variables* (Springer, New York, 1986)
36. R.M. Range, WHAT IS… a pseudoconvex domain? Not. Am. Math. Soc. **59**(2), 301–303 (2012)
37. R.C. Reilly, Applications of the Hessian operator in a Riemann manifold. Indiana Univ. Math. J. **26**, 459–472 (1977)
38. J. Serrin, A symmetry problem in potential theory. Arch. Ration. Mech. Anal. **43**, 304–318 (1971)
39. Z. Slodkowski, G. Tomassini, Weak solutions for the Levi equation and envelope of holomorphy. J. Funct. Anal. **101**, 392–407 (1991)
40. Z. Slodkowski, G. Tomassini, The Levi equation in higher dimensions and relationships to the envelope of holomorphy. Am. J. Math. **116**(2), 479–499 (1994)

41. W. Süss, Uber Kennzeichnungen der Kugel und Affinesphären durch Herrn K.P. Grotemeyer. Arch. Math **3**, 311–313 (1952)
42. G. Tomassini, Geometric properties of solutions of the Levi equation. Ann. Mat. Pura Appl. (4) **152**, 331–344 (1988)
43. N.S. Trudinger, Isoperimetric inequalities for quermassintegrals. Ann. Inst. H. Poincaré Anal. Non Linéaire **11**(4), 411–425 (1994)
44. J.I.E. Urbas, On the existence of nonclassical solutions for two classes of fully nonlinear elliptic equations. Indiana Univ. Math. J. **39**(2), 355–382 (1990)

List of Participants

1. Abbondanza Beatrice
   beatrice.abbondanza@hotmail.it
2. Capogna Luca
   lcapogna@ima.umn.edu
3. Caramuta Pietro
   pietro.caramuta
4. Chinni Gregorio
   gregorio.chinni3@unibo.it
5. Cupini Giovanni
   giovanni.cupini@unibo.it
6. Darvas Tamas
   tdarvas@math.purdue.edu
7. Dolce Emanuela
   emanela.dolce@alice.it
8. Dragomir Sorin
   sorin.dragomir@unibas.it
9. Fanciullo Maria Stella
   fanciullo@dmi.unict.it
10. Gu Dongwei
    gudw09@gmail.com
11. Guan Pengfei
    guan@math.mcgill.ca
12. Gutierrez Cristian
    gutierre@temple.edu
13. Katzourakis Nikolaos
    nkatzourakis@bcamath.rg
14. Kim Ha Ly
    lykimha@math.unipd.it
15. Kogoj Alessia Elisabetta
    alessia.kogoj@unibo.it

L. Capogna et al., *Fully Nonlinear PDEs in Real and Complex Geometry and Optics*,
Lecture Notes in Mathematics 2087, DOI 10.1007/978-3-319-00942-1,
© Springer International Publishing Switzerland 2014

16. Lanconelli Ermanno
    lanconel@dm.unibo.it
17. Martino Vittorio
    vmartino@sissa.it
18. Montanari Annamaria
    annamaria.montagnani@unibo.it
19. Sabra Ahmad
    tuc70013@temple.edu
20. Salani Paolo
    salani@math.unifi.it
21. Salort Ariel
    asalort@dm.uba.ar
22. Scarola Cristian
    cristian.scarola@gmail.com
23. Tralli Giulio
    giulio.tralli2@unibo.it
24. Vespri Vincenzo
    vespri@math.unifi.it

LECTURE NOTES IN MATHEMATICS  Springer

Edited by J.-M. Morel, B. Teissier; P.K. Maini

Editorial Policy (for Multi-Author Publications: Summer Schools / Intensive Courses)

1. Lecture Notes aim to report new developments in all areas of mathematics and their applications - quickly, informally and at a high level. Mathematical texts analysing new developments in modelling and numerical simulation are welcome. Manuscripts should be reasonably selfcontained and rounded off. Thus they may, and often will, present not only results of the author but also related work by other people. They should provide sufficient motivation, examples and applications. There should also be an introduction making the text comprehensible to a wider audience. This clearly distinguishes Lecture Notes from journal articles or technical reports which normally are very concise. Articles intended for a journal but too long to be accepted by most journals, usually do not have this "lecture notes" character.

2. In general SUMMER SCHOOLS and other similar INTENSIVE COURSES are held to present mathematical topics that are close to the frontiers of recent research to an audience at the beginning or intermediate graduate level, who may want to continue with this area of work, for a thesis or later. This makes demands on the didactic aspects of the presentation. Because the subjects of such schools are advanced, there often exists no textbook, and so ideally, the publication resulting from such a school could be a first approximation to such a textbook. Usually several authors are involved in the writing, so it is not always simple to obtain a unified approach to the presentation.

   For prospective publication in LNM, the resulting manuscript should not be just a collection of course notes, each of which has been developed by an individual author with little or no coordination with the others, and with little or no common concept. The subject matter should dictate the structure of the book, and the authorship of each part or chapter should take secondary importance. Of course the choice of authors is crucial to the quality of the material at the school and in the book, and the intention here is not to belittle their impact, but simply to say that the book should be planned to be written by these authors jointly, and not just assembled as a result of what these authors happen to submit.

   This represents considerable preparatory work (as it is imperative to ensure that the authors know these criteria before they invest work on a manuscript), and also considerable editing work afterwards, to get the book into final shape. Still it is the form that holds the most promise of a successful book that will be used by its intended audience, rather than yet another volume of proceedings for the library shelf.

3. Manuscripts should be submitted either online at www.editorialmanager.com/lnm/ to Springer's mathematics editorial, or to one of the series editors. Volume editors are expected to arrange for the refereeing, to the usual scientific standards, of the individual contributions. If the resulting reports can be forwarded to us (series editors or Springer) this is very helpful. If no reports are forwarded or if other questions remain unclear in respect of homogeneity etc, the series editors may wish to consult external referees for an overall evaluation of the volume. A final decision to publish can be made only on the basis of the complete manuscript; however a preliminary decision can be based on a pre-final or incomplete manuscript. The strict minimum amount of material that will be considered should include a detailed outline describing the planned contents of each chapter.

   Volume editors and authors should be aware that incomplete or insufficiently close to final manuscripts almost always result in longer evaluation times. They should also be aware that parallel submission of their manuscript to another publisher while under consideration for LNM will in general lead to immediate rejection.

4. Manuscripts should in general be submitted in English. Final manuscripts should contain at least 100 pages of mathematical text and should always include

   – a general table of contents;
   – an informative introduction, with adequate motivation and perhaps some historical remarks: it should be accessible to a reader not intimately familiar with the topic treated;
   – a global subject index: as a rule this is genuinely helpful for the reader.

   Lecture Notes volumes are, as a rule, printed digitally from the authors' files. We strongly recommend that all contributions in a volume be written in the same LaTeX version, preferably LaTeX2e. To ensure best results, authors are asked to use the LaTeX2e style files available from Springer's web-server at
   ftp://ftp.springer.de/pub/tex/latex/svmonot1/ (for monographs) and
   ftp://ftp.springer.de/pub/tex/latex/svmultt1/ (for summer schools/tutorials).
   Additional technical instructions, if necessary, are available on request from:
   lnm@springer.com.

5. Careful preparation of the manuscripts will help keep production time short besides ensuring satisfactory appearance of the finished book in print and online. After acceptance of the manuscript authors will be asked to prepare the final LaTeX source files and also the corresponding dvi-, pdf- or zipped ps-file. The LaTeX source files are essential for producing the full-text online version of the book. For the existing online volumes of LNM see:
   http://www.springerlink.com/openurl.asp?genre=journal&issn=0075-8434.
   The actual production of a Lecture Notes volume takes approximately 12 weeks.

6. Volume editors receive a total of 50 free copies of their volume to be shared with the authors, but no royalties. They and the authors are entitled to a discount of 33.3 % on the price of Springer books purchased for their personal use, if ordering directly from Springer.

7. Commitment to publish is made by letter of intent rather than by signing a formal contract. Springer-Verlag secures the copyright for each volume. Authors are free to reuse material contained in their LNM volumes in later publications: a brief written (or e-mail) request for formal permission is sufficient.

Addresses:
Professor J.-M. Morel, CMLA,
École Normale Supérieure de Cachan,
61 Avenue du Président Wilson, 94235 Cachan Cedex, France
E-mail: morel@cmla.ens-cachan.fr

Professor B. Teissier, Institut Mathématique de Jussieu,
UMR 7586 du CNRS, Équipe "Géométrie et Dynamique",
175 rue du Chevaleret, 75013 Paris, France
E-mail: teissier@math.jussieu.fr

For the "Mathematical Biosciences Subseries" of LNM:

Professor P. K. Maini, Center for Mathematical Biology,
Mathematical Institute, 24-29 St Giles,
Oxford OX1 3LP, UK
E-mail: maini@maths.ox.ac.uk

Springer, Mathematics Editorial I,
Tiergartenstr. 17,
69121 Heidelberg, Germany,
Tel.: +49 (6221) 4876-8259
Fax: +49 (6221) 4876-8259
E-mail: lnm@springer.com